The Sustainable Renovation of Buildings and Neighbourhoods

Edited By

Pilar Mercader-Moyano

Department of Building Constructions I
University of Seville
Spain

CONTENTS

FOREWORD

The eco-efficient refurbishment of buildings and sets of building forming neighbourhoods have become a new paradigm to the building agents. It seems that the numerous messages which constantly reiterate the need to go deeper into the development of Eco-Efficient solutions, to make the development being sustainable, have taken hold in the new generations of researchers and technicians.

One of the fruits of this new situation is the II International Congress on Sustainable Construction and Eco-Efficient Solutions, which has as a precedent four national editions, attending researchers and technicians, stimulated by the continuous messages received from all the angles, and who are convened to these forums to show their ideas.

The prestige achieved by the Congress on Sustainable Construction, by the quality of the communications presented on previous occasions, has driven to this edition's Scientific-Technical Committee to make an extraordinary effort to select the best papers presented, through a blind-peer review process, thereby elevating the scientific quality of the Congress.

The great quality of the communications presented has allowed to highlight six of an excellent value, linked to the Eco-Efficient Buildings and Neighbourhoods Renovations.

CHAPTER 1: Energy Retrofitting for Social Housing by Improving the Building Envelope: Madrid 1939-1979.

This piece of work is part of a study about the improving possibilities in the constructive systems of vertical exterior walls in social housing in Spain, with two parallel objectives: to improve the environmental quality inside the dwellings, and to improve its energy efficiency.

The increase in the environmental requirements for buildings, and particularly with the requirements of energy savings and efficiency, makes necessary the study of the improvement potential of the different construction systems. A good façade design, adapted to local constraints, such as climate or solar access, is key to guarantee the habitability, as well as to reduce the environmental impacts.

Actions in vertical exterior walls, apart from improving the energy performance, will allow to renew and improve other kinds of necessities, such as the image of the building, with a consequent positive impact at a neighbourhood scale. The implementation of energy improvements can also be sustained in other necessary actions, like maintenance works or repair works, so the costs are optimized to different actions.

The improvement of the dwellings habitability is sought as an objective, as well as the reduction of the environmental impacts in the life cycle of the construction systems. In both cases, the façade plays a leading role.

CHAPTER 2: Environmental Impact Indicator System in Residential Neighbourhood Renovation.

There is a specific urban tissue in Spanish cities which is the open-block neighbourhoods, built from the fifties to the end of the seventies, based on open building typologies. These neighbourhoods, or “barriadas”, have inherited a state resulted from the important quantitative needs of existing housing at the end of the wars of the mid twentieth century: work processes within the short execution, reduced costs, poor and emerging industrialized production and comfort standards completely away from today’s minimum required demands.

Within these neighbourhoods, the research focuses on those whose design is the result of the repetition of one or more residential typologies of collective housing. As a response to this situation, the research proposes the evaluation and minimization of the environmental impact produced by these fabrics in the cities, to identify the key areas that contribute to make their environmental rehabilitation.

This magnificent work proposes an indicator system which focuses on these areas responsible for the imbalances produced in the urban ecosystems, which is based on an exhaustive study of the current system, to the urban level and also to the building level. As a result of the study, a new indicator system is developed, being adapted to this specific reality.

CHAPTER 3: Approach to the Definition of Nearly Zero Energy Buildings, in the Social Housing in Europe.

This chapter presents the results of the Power House NZEB Challenge project. The primary objective of the project is to transfer and exchange experiences between different countries about the design of types of buildings, and inform to political lawmakers about the results. The project is studying thirty different buildings by following the optimal cost methodology. This chapter presents one of the study cases from Spain, showing the results of applying the methodology of the optimal cost, which allows to identify those technical solutions with the higher economical profitability considering the global cost of the next 30 years. The knowledge generated will help to define the NZEB concept and will be useful as a guide for the Member States on the elaboration of the regulatory and financial framework; to assure that energy transition is inclusive and socially, economically and environmentally sustainable.

CHAPTER 4: Development of an Assessment Tool for Building Envelope Retrofit Based on Environmental Indicators.

In this work, it is confirmed that we are attending a wide change in the sector, which aims to achieve a cultural change which transforms the repair objective into a deeper one about sustainable management of the buildings, in which the active prevention, conservation and maintenance improve the quality of life of their inhabitants and reduce their use costs.

To that end, there are proposed tools that provide reliably significant indicators related to the building sustainability. Thus, the main objective of this work is to define an evaluation model of the environmental impact related to the energy rehabilitation of the building envelopes, to be able to assign a rating by recognizable indicators of sustainability.

The evaluation model defined allows to assign values to the indicators: CO2 emissions, energy consumption, drinking water, generation of dangerous waste and non-dangerous and maintenance costs. Besides, it provides a scale that allows to relate the impacts in all the phases of the building life cycle.

CHAPTER 5: A structure for the Quantity Surveillance of Costs and Environmental Impact of Cleaning and Maintenance in Buildings.

The pause resulted by the paralysis of the sector gives an opportunity to reflex the considerable size of the costs produced by the maintenance of the buildings constructed.

Among the maintenance labours, the cleaning leads from the rest of the volume of work and frequency it requires. Although it has also been a great impact in residential and public buildings, the cleaning takes on maximum importance in hotels, where it becomes an essential work, on which depends the progress of the business.

In this area, the frequency with which each activity is performed grows exponentially, leading us to consider: how much they cost and how much environmental impact these works generate along the life cycle of the building?

To solve this problem, in this paper, price development is proposed which compiles the costs and environmental impact of the cleaning works, being the costs based on the guidelines of the Base of Construction Costs of Andalusia (Base de Costes de la Construcción de Andalucía BCAA). To that end, a new chapter is written dedicated to the phase of use and maintenance of buildings, which presents a continuity with the existing Classification System.

Among the prices developed, on the one hand, they distinguished the cleaning of specific elements, and on the other hand, the one of the different kind of stay with a specific periodicity. In that way, when the measure is run, the initial ridiculous costs become considerable amounts of money due to the large number of repetitions.

As an immediate result of this work, the foundations have been laid for the creation of a new chapter of the BCCA with a solid structure.

CHAPTER 6: Management of Living Urban Environments from an Active Ageing Approach. Advances in the Team Research (Re)Programa.

In the near future, we will find a socioeconomic situation with a difficult solution: our city's population and their neighbourhoods will be mainly formed by elderly people, that will live mostly in obsolete dwellings, non-adapted to their housing requirements, with the aggravating factor that they cannot count on the economic capacity to be able to cover the rehabilitation costs of the building where they live.

In this research the objectives, criteria, analysis methodology, evaluation tools and case studies are exposed and analysed by the research team Reprograma in the project entitled "(Re)habitation+ (Re) generation+ (Re)programming. Renovation and sustainable management of the Andalusian building stock. Management of living urban environments from active ageing, gender and habitability approach" funded by the Andalusia Government´s Public Works Agency, between 2013 and 2015. It also shared a result advance, relating the information generated to date.

CHAPTER 7: Energy Efficiency Indicators for Buildings in Argentina.

In this work, a comparative study of energy performance in different types of buildings is made. It analysed three cases of office and three of residential buildings, which are compared with samples of previous researches, looking for correlative or different behaviour in terms of potential energy losses and gains. Then, the indicators and indexes are compared with Argentinean Standards to verify their efficiency level by using the "Energo-CAD" methodology and energy model, based on steady-state balances. While discussing the results of energy audits that show discomfort with high energy demand, which are related to energy breakdown in this country, it is concluded that the indicators employed show clear results, and facilitate the comparison among studied cases. This work ends showing that the sample does not meet the National Standards, even when it is compulsory by law since 2003 in the province of Buenos Aires.

CHAPTER 8: Towards a Sustainable Retrofitting Plan for Social Housing in Mediterranean Europe.

The research is focused on a characteristically Mediterranean city in the south of Spain, Cadiz, in climate zone A3 according to its legislation and regulation context. This research aims to cover the lack of information on the energy behaviour of the existing housing stock, particularly in social housing neighbourhoods built prior to regulations (1940-1980) and the establishment of a Plan for Sustainable Retrofitting of social housing in the city of Cadiz.

In order to implement this plan, the chapter identifies and classifies the main social housing neighbourhoods built in that period, offering a construction and energy analysis of the constructive systems used in the thermal envelope, which has made it possible to classify and establish their degree of obsolescence. This study has provided a global assessment of the energy rating of social housing built in the city in 1940-1980, applied to a map of the city allowing us to view an energy rating map of sorts of the city. This method is essential in order to attain what could be described as an essential first phase towards attaining the priority action goals of Horizon 2020, using global actions with sustainable retrofitting plans for Spanish cities.

CHAPTER 9: New Applications of Ceramic Waste Generated in Building Rehabilitation Works.

This chapter presents the results of a research project conducted to analyze the feasibility of incorporating ceramic brick and tile waste from construction in a plaster matrix, through a study carried out in laboratory and ensuring meeting the current regulations. Results show that it is possible to incorporate up to 25% of ceramic waste over the weight of gypsum without changing its basic properties. Perforated brick waste achieves the best behavior: increases mechanical strengths (31% compression and 29% flexural) and surface hardness (8%) and reduces water absorption around 25%.

Antonio Ramírez-de-Arellano-Agudo
Department of Building Constructions II
University of Seville
Spain

PREFACE

The chapters published in this book are taken from the II International Congress on Sustainable Construction and Eco-Efficient Solutions, held in Seville, Spain, in May 2015, this being the fourth edition at a national level. From the 150 papers presented, the best papers have been selected in a double-blind peer review by an International Scientific-Technical Committee. These papers are collected in this book.

This Congress edition, organized by the University of Seville and supported by the Department of Development and Housing of the Government of Andalusia (Consejería de Fomento y Vivienda de la Junta de Andalucía), has the motto: "Eco-efficient refurbishment of buildings and neighbourhoods".

Refurbishing building constructions in a more sustainable way, by minimizing the environmental impact, is the objective sought, in response not only to the European approach Horizon 2020, but to all those who seek a more sustainable world. This objective has brought together more than 100 international researchers, representatives from Argentina, Chile, Ecuador, Mexico, Portugal, and Spain and 200 students from the branches of Architecture, Engineering, and Environmental Sciences, among others.

Throughout three days, professionals, scientists, researchers, and public administration representatives worked together in thematic and parallel blocks, in round tables and in debating sessions, in order to reflect on the decision-making that helps to improve the technique innovation in both the public and private building sectors.

At the same time, the students worked intensively in the workshops to achieve innovative building solutions, as alternatives to conventional solutions, which are capable of minimising the environmental impact generated by construction, thereby improving the energy efficiency of the pre-existent buildings, with an optimal economic cost, and affordable in the long-term.

The chapters of this book provide a summary of the main debates, papers and the results discussed in the referenced sessions.

We wish to thank all the conference participants and especially the staff of the organisation of the Congress for their valuable contributions.

Pilar Mercader-Moyano
Department of Building Constructions I
University of Seville
Spain

List of Contributors

Alberto Vilches	University of Seville, Department of Building Construction I, Reina Mercedes 2 Avenue, 41012, Seville, Spain
Alejandra García-Prieto Ruiz	IVE Valencia Institute of Building, Retrofit Department, C/ Tres Forques, 98, 46018, Valencia, Spain
Alejandro Martínez-Rocamora	Department of Building Constructions II, ETSIE, University of Seville, Av. Reina Mercedes, 4-a. 41012 – Seville, Spain
Ángela Barrios	University of Seville, Department of Building Construction I, Reina Mercedes 2 Avenue, 41012, Seville, Spain
Antonio Ramírez-de-Arellano	University of Seville, Department of Building Constructions II, Reina Mercedes 2 Avenue, 41012, Seville, Spain
Antonio Serrano	University of Seville, Department of Building Construction I, Reina Mercedes 2 Avenue, 41012, Seville, Spain
Begoña Serrano Lanzarote	IVE Valencia Institute of Building, Retrofit Department, C/ Tres Forques, 98, 46018, Valencia, Spain
Belinda López-Mesa	Universitat de Zaragoza, Spain
Carmen Alonso	Department of Construction, Eduardo Torroja Institute of Construction Sciences, Spanish National Research Council, Madrid, Spain
Elena Mulet	Universitat Jaume I, Spain
Fernando Martín-Consuegra	Department of Construction, Eduardo Torroja Institute of Construction Sciences, Spanish National Research Council, Madrid, Spain
Francisco J. Lizana	University of Seville, Department of Building Construction I, Reina Mercedes 2 Avenue, 41012, Seville, Spain
Ignacio Oteiza	Department of Construction, Eduardo Torroja Institute of Construction Sciences, Spanish National Research Council, Madrid, Spain
Jaime Solís-Guzmán	Department of Building Constructions II, ETSIE, University of Seville, Av. Reina Mercedes, 4-a. 41012 – Seville, Spain
Jorge D. Czajkowski	Laboratory of Sustainable Architecture and Habitat, Faculty of Architecture and Urbanism, National University of La Plata, CONICET, 47 St. no.162- 1900 La Plata – Buenos Aires Province- Argentina

José C. Mariñas	University of Seville, Department of Building Construction I, Reina Mercedes 2 Avenue, 41012, Seville, Spain
Juan J. Sendra Salas	Instituto Universitario de Arquitectura y Ciencias de la Construcción, Escuela Técnica Superior de Arquitectura, Universidad de Sevilla, Spain
Juan Monjo	Department of Construction and Architectural Technology, Higher Technical School of Architecture, Polytechnic University of Madrid, Spain
Julia Garrido-Piñero	University of Seville, Department of Building Constructions I, Reina Mercedes 2 Avenue, 41012, Seville, Spain
Laura Soto Francés	IVE Valencia Institute of Building, Retrofit Department, C/ Tres Forques, 98, 46018, Valencia, Spain
Leticia Ortega Madrigal	IVE Valencia Institute of Building, Retrofit Department, C/ Tres Forques, 98, 46018, Valencia, Spain
Madelyn Marrero	Department of Building Constructions II, ETSIE, University of Seville, Av. Reina Mercedes, 4-a. 41012 – Seville, Spain
Mariam González-Moya	Department of Construction and Architectural Technology, Higher Technical School of Architecture, Polytechnic University of Madrid, Spain
Mariano González Cortina	Escuela Técnica Superior de Edificacion (UPM), Department of Building Construction and its Control, Avenida Juan de Herrera, 6 28040 Madrid, Spain
Marta Molina	University of Seville, Department of Building Structures and Geotechnical Engineering, Reina Mercedes 2 Avenue, 41012, Seville, Spain
Mercedes Del Río Merino	Escuela Técnica Superior de Edificación (UPM), Department of Building Construction and its Control, Avenida Juan de Herrera, 6 28040 Madrid, Spain
Mónica Martínez Hervás	Instituto Universitario de Arquitectura y Ciencias de la Construcción, Escuela Técnica Superior de Arquitectura, Universidad de Sevilla, Spain
Paola Villoria Saéz	Escuela Técnica Superior de Edificación (UPM), Department of Building Construction and its Control, Avenida Juan de Herrera, 6 28040 Madrid, Spain
Patricia Huedo	Universitat Jaume I, Spain
Pilar Mercader-Moyano	University of Seville, Department of Building Constructions I, Reina Mercedes 2 Avenue, 41012, Seville, Spain

Rafael Suárez Medina Instituto Universitario de Arquitectura y Ciencias de la Construcción, Escuela Técnica Superior de Arquitectura, Universidad de Sevilla, Spain

Rocío Santos Jiménez Escuela Técnica Superior de Edificación (UPM), Department of Building Construction and its Control, Avenida Juan de Herrera, 6 28040 Madrid, Spain

Violeta de la Fuente Pérez IVE Valencia Institute of Building, Retrofit Department, C/ Tres Forques, 98, 46018, Valencia, Spain

CHAPTER 1

Energy Retrofitting for Social Housing by Improving the Building Envelope: Madrid, 1939-1979

Ignacio Oteiza[1,*], Carmen Alonso[1], Fernando Martín-Consuegra[1], Juan Monjo[2] and Mariam González-Moya[2]

[1]Department of Construction, Eduardo Torroja Institute of Construction Sciences, Spanish National Research Council, Madrid, Spain and [2]Department of Construction and Architectural Technology, Higher Technical School of Architecture, Polytechnic University of Madrid, Spain

Abstract: This work forms part of a study into the possibilities of improving the construction systems used for exterior enclosing walls of social housing in Spain. It has two parallel objectives, to improve the quality of the environment inside the houses and to enhance their energy efficiency. The increase in environmental demands placed on buildings, particularly in terms of energy saving and efficiency, has led to the need to study potential improvements to the different construction systems currently available. A well-designed facade adapted to local conditions, such as climate or solar access, is a key factor in guaranteeing the building's habitability, while also playing an important role in decreasing environmental impacts. Retrofits on exterior enclosing walls, besides improving energy performance, also afford the opportunity to renovate and improve other kinds of needs, *e.g.*, the building's image, producing a positive impact on a neighbourhood-wide scale. Energy efficiency improvements can also be deployed in conjunction with other activities that may be necessary, such as maintenance or repair work, thereby optimising the costs of the different interventions. The objectives proposed herein are firstly to improve the habitability of the homes and, secondly, to reduce the environmental impact of the construction systems within their life-cycle. The facade plays a key role in achieving both of these aims.

Keywords: Building envelope, building stock regeneration, construction systems, energy efficiency, energy poverty, energy retrofit, energy consumption, façade, madrid: 1939-1979, materials, passive measures, social housing, thermal behavior, thermal transmittance.

INTRODUCTION

The aim of this work is to study the construction of opaque exterior vertical enclosing walls in buildings containing multi-family dwellings constructed in the

***Corresponding author Ignacio Oteiza:** Department of Construction, Eduardo Torroja Institute of Construction Sciences, Spanish National Research Council, Madrid, Spain; Tel: +34 913020440; Fax: +34 913020700; E-mail ioteiza@ietcc.csic.es

Pilar Mercader-Moyano (Ed.)

period between the end of the Spanish Civil War (1939), which was immediately followed by migration of rural populations to large cities and a subsequent increase in housing construction activity, until the appearance of the first standard to address insulation in opaque vertical enclosures, standard NBE-CT-79 [1]. This period coincides, apart from the last four years, with Franco's dictatorship in Spain.

The purpose is to understand the current situation of these homes and their envelopes through different factors that play a role in energy retrofitting, mainly in terms of the construction, but without forgetting other essential aspects, such as economic or social factors, that afford the opportunity to tackle the matter from the perspective of sustainability, which is particularly relevant when studying practical solutions for improving social housing.

There is no doubt about the potential energy savings that could be achieved by retrofitting housing constructed between 1939 and 1979, especially if the policies focus specifically on improving energy efficiency. Pressure comes from Europe through standards and policies contained within the strategies for Horizon 2020, such as Directive 2012/27/EU and its application in Spain.

SOCIAL HOUSING

This point deals with some of the criteria that define the social housing addressed by this study. Social housing is understood to be all housing developments promoted by the state, or by a private entity but which receives assistance in the form of financial support or tax concessions granted by the state. In Spain, it is also known as officially or publicly promoted housing. We explain some of the indicators and concepts applied in previous studies of Spanish cases by this and other research groups; said studies are essentially recent investigations into energy retrofitting for housing.

Background

Social housing in Europe arose from the need to accommodate ever-increasing urban populations produced by rural-urban migrations which began in the middle of the 19th century. Several central European countries can provide examples of important social housing developments, particularly noteworthy are the projects in Germany (see the case of Frankfurt between 1890 and 1914) and Austria (throughout the 1920s) [2]. Social housing projects in Spain also began to emerge around the end of the 19th and beginning of the 20th centuries, with the enactment of special laws such as the “Law for cheap accommodation” in 1908 and the

"Law for cheap houses" of 1911. However, the present work concentrates on social housing constructed in Madrid in the period between the end of the Spanish Civil War (1939) and the 1979 introduction of the first standard (NBE-CT-79) to include thermal criteria.

Madrid 1939 and 1979

After the Spanish Civil War, there was generally a very large requirement for housing and infrastructure but it was inversely proportional to the available resources and so, therefore, no significant housing developments appeared until the 1950s. When Moya discussed social housing in the book "*Barrios de promoción oficial 1939-1976*" ["Officially promoted neighbourhoods 1939-1976"] [2], he distinguished two large, clearly differentiated periods, the first between the years 1939 and 1961 (A) and a second phase from 1961 to 1976 (B). During the first period, characterised by Spain's autocracy up to 1961, a large number of houses were constructed according to the criterion of quantity above quality. Sambricio [3] asserts that "social housing" in the years just after the war was only destined for civil servants, military personnel or certain labourers from the trade union aligned to the regime.

The first housing plan was enacted in 1944 and proposed the construction of 1.4 million houses; 360,000 to address the existing shortfall, another 400,000 for purposes of relocation and 640,000 to cover the increasing population. A second housing plan was introduced in 1955. Housing developments in Madrid were supported through various government bodies, such as the Trade Union Homebuilding Initiative (OSH), Madrid City Council, Urban Planning Commission for Madrid and its surroundings, as well as housing trusts established for ministries and companies controlled by the National Institute for Industry (INI). There was an urgent need to construct houses in early post-Civil War Spain, yet the available resources were not enough to meet the demand and there was an excess of workers but no means of production to put them to use. At the end of the First World War, countries such as Germany, France and England resumed their paths toward the industrialisation of previously standardised, prefabricated construction elements. In 1949, the Torroja Institute held an international competition for the industrialised construction of 50,000 houses. There was massive international participation, but the competition was shelved [4]. Several further competitions were organised during the 1950s. One of those, in December 1954, was directed towards constructors associated with architects. This was also the period when quality labels were first applied to construction.

Due to the large number of government agencies promoting housing developments, in 1957 an attempt was made to centralise housing policies by creating the Ministry of Housing. Nevertheless, housing continued to be a significant problem in the capital and so the "*Madrid social emergency plan*" was approved. This included the "*Managed towns*" (1957) (Canillas, Caño Roto, Manoteras, Fuencarral, San Cristóbal, San Blas, La Elipa, amongst others) and the "*Small and absorbed towns*" in 1958.

"Bidagor's 1941 *Cities Plan* produced the first proposals for reconstructing post-war Madrid (Fig. **1**). It suggested hierarchical decentralisation, articulated through the concepts of: neighbourhood, district and urban centre. The plan proposed cohesive growth for Madrid, adapted to topography and with a series of green concentric rings that would also accommodate the circulation of traffic. However, the sluggish execution of the project meant that the green spaces were spontaneously occupied by quickly erected buildings, thus collapsing the periphery. In light of the situation, a model was adopted that involved creating new satellites from scratch" [5].

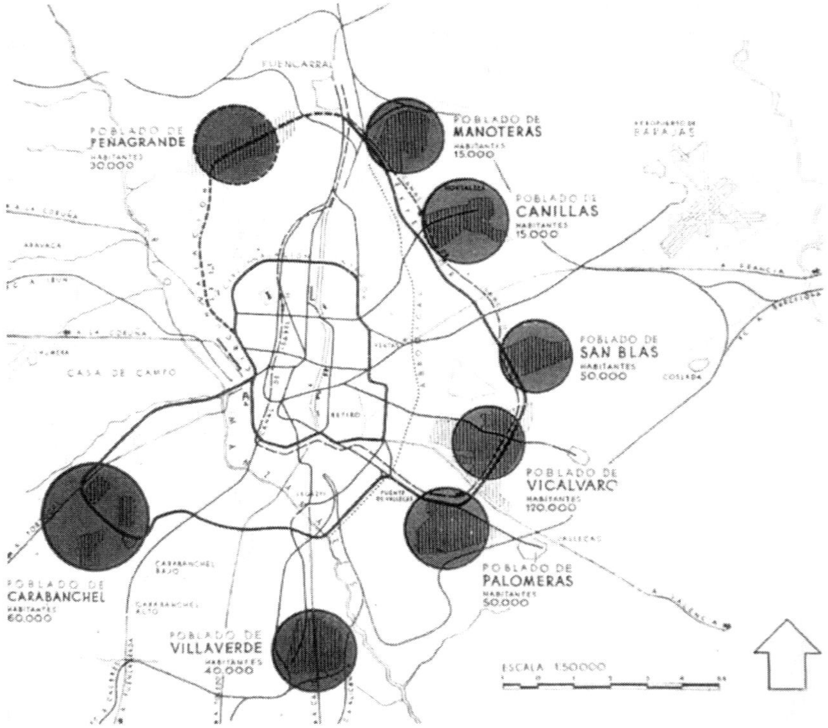

Figure 1: Bidagor, 1941 Cities Plan.

The second period (B), between 1961 and 1979, saw the beginnings of large economic changes in Spain which would have a direct influence on housing policy. The number of houses generated by public bodies was reduced and private development projects were given greater impetus. This lead to a new period of reflection regarding urban and construction typologies, although the problematic suburbs and shanty towns were still present at the start of this phase. "Around 20% of housing was promoted publicly, 30% was developed privately but with public support and 50% was for private development" [2]. A large quantity of houses were produced in the 1970s, principally through private development. The transition[1] encountered a situation with an enormous number of houses from the 1950s that were in a very bad state, in some cases beyond repair, giving rise to a first phase of neighbourhood remodelling between 1975-80.

Fig. **2** represents a series of milestones considered to be important for characterising current stock in the Municipality of Madrid, it highlights: the number of first, second and vacant homes in primarily residential buildings according to the 2011 population and housing census; the end of the Civil War (1936-39); the buildings present before 1951, as an indicator of urban vulnerability; the start of the successive housing plans, in particular the III National Housing Plan during the dictatorship (1961-76) and the major neighbourhood remodelling schemes, specifically those carried out in Madrid by the local housing institute, *IVIMA*, from the 1970s onwards; the oil crisis, followed by the 1979 introduction of energy efficiency criteria in building regulations; the peak of the real estate boom and the subsequent economic crisis it triggered when the bubble burst in 2006; the publication of Spain's Technical Building Code, *CTE*, in 2006; and also marked is the transposition of various, recently published European plans and directives that address energy efficiency and set out clear short- and long-term objectives regarding the reduction of greenhouse gas emissions [6].

The two primary periods of interest for this study are also outlined on the figure, A and B, 1939-60 and 1961-79 respectively, demonstrating obvious differences in terms of the number of homes; their construction and energy characteristics are described below. Large-scale migrations to urban centres plus a lack of resources created a residential building stock marked by the successive plans and developments carried out during the dictatorship. In the seventies, neighbourhood organisations completed various neighbourhood remodelling plans which improved quality standards and generally addressed the different types of

[1] The period of political transition from a dictatorship to a democracy following the death of Franco in 1975.

substandard housing settlements. In Madrid these could be classed as: neighbourhoods/towns composed of large concentrations of makeshift housing, small shanty enclaves, ethnic minority settlements (mainly Gypsy) and areas of ramshackle public housing constructed by the National Housing Institute and the Trade Union Homebuilding Initiative.

In many of the world's developed countries, residential buildings account for a significant portion of the general building stock's energy consumption. In Europe the building sector represents up to 40% of total energy consumption [6], but we should not forget that the transport sector is the largest consumer, with disperse cities requiring more energy than compact ones.

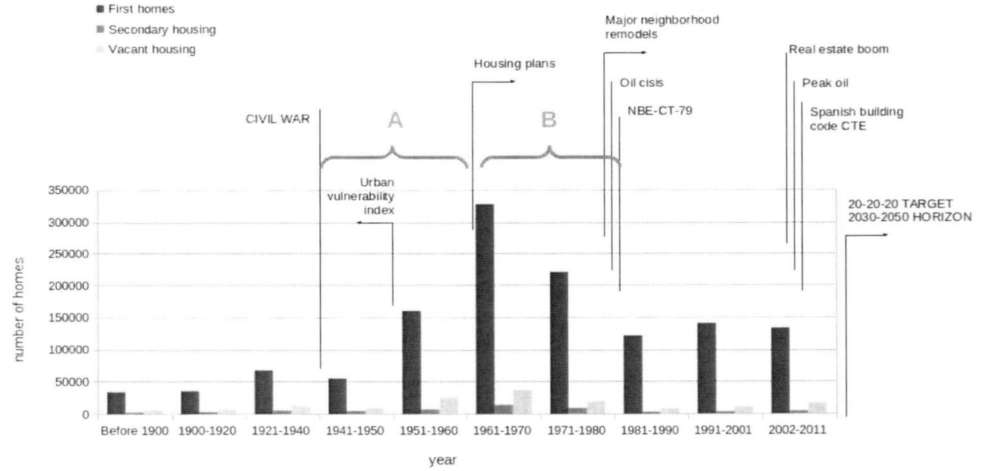

Figure 2: Number of first, second and vacant homes in the Municipality of Madrid according to the year of construction. (Data from the Spanish National Institute of Statistics, *INE*.) [7].

According to the 2011 census [7], residential building stock totalled over 25 million units, 42% of which were constructed between 1940 and 1980 [8], if we add this figure to the houses built before 1939 then it represents 61% of all houses, *i.e.*, over 15 million homes were built without following any thermal criteria and only 39% of current housing was constructed after the introduction of standard NBE-CT-79 (Fig. **3**).

If one analyses the building stock according to construction year, as shown in Fig. **3** for the case of Madrid in relation to the Spanish average, then the final period from the year 2000 onwards is notable as it reveals that the real estate boom was responsible for over 25% of present housing stock; stock which coincides with the current reference standard, the DB-HE contained within the Spanish Technical Building Code, *CTE* [9].

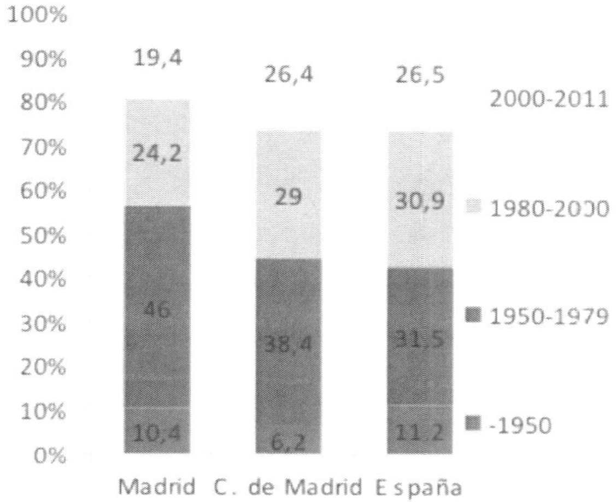

Figure 3: Percentage of residential buildings according to their year of construction. (Compiled from data provided by the *INE* of the Ministry of Internal Development).

Moving back in time, the next regulatory milestone was the publication of standard NBE-CT-79, the first to incorporate standards for thermal insulation in the facades and which now affects approximately one third of current building stock.

This study has investigated social multi-family residential buildings for the following reasons: 1) they are very common and representative of large cities, 2) due to their socio-economic element, and 3) because the surface of the facade has a significant impact with respect to the total surface area of the envelope.

The Need for Energy Retrofitting

A large portion of the social residential buildings constructed between 1939 and 1979 have been reformed and partially adapted to some of the regulations that have appeared over the years (Fig. **4**). Nevertheless, many of them still suffer significant pathological processes of deterioration, especially affecting the building envelope and the public spaces where they are located, primarily due to the poor quality materials used in periods of limited economic resources. The majority of the residential building stock presents a state of deterioration that requires investment to rejuvenate the buildings and neighbourhoods. The neighbourhoods and their buildings definitely require active intervention to bring them up to current standards of comfort, both in terms of accessibility and energy efficiency. Improving the building envelope not only helps achieve energy efficiency and higher levels of comfort, but it can also be used as a supporting tool or complement to resolve different types of construction damage or problems of functional suitability (noise, accessibility, *etc.*) and especially for improving the exterior spaces that are subject to the social problems of these often degraded areas of the city.

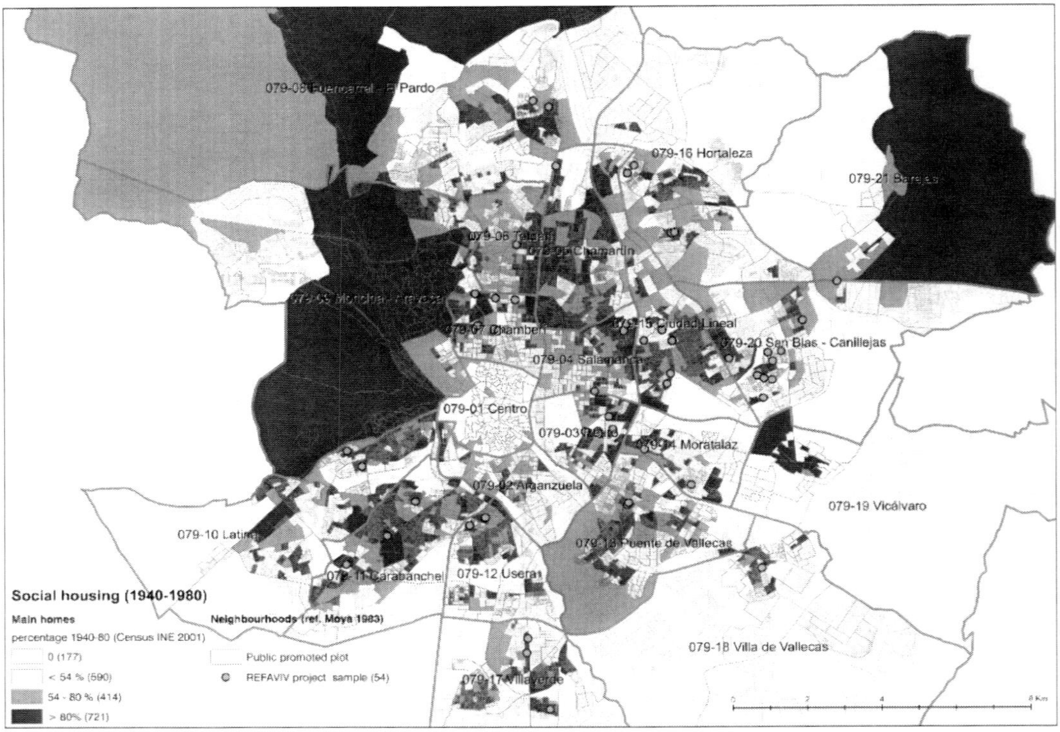

Figure 4: Plan of present day Madrid, percentage of first houses (homes) per censual zone constructed during the study period of the *REFAVIV* project (1940-80) with respect to the average for the Municipality of Madrid (54%); data taken from the *INE* Census 2001. Officially protected housing (*VPO*) is marked in red [2]. (Compiled by the author).

The 2014 report by the Restoration Working Group [10] indicates that, "the RWG's assessment reveals the significant potential for the renovation of Spanish buildings, while it also indicates that the legislative, financial and operational frameworks in place are insufficient to transform the sector and yield economic, social and environmental advantages. We have identified some 10 million first homes and several 'sectorial strategies' among the commercial buildings. These strategies can transform the energy and environmental footprint of Spanish buildings, and also create around 150,000 sustainable, long-term jobs in a New Building Sector. However, the necessary legislative, financial and operational conditions required to initiate this process do not yet exist in Spain. It is worth highlighting that a multi-level strategy is required which involves all the different agents in the sector".

THE ENERGY INEFFICIENCY OF RESIDENTIAL BUILDING STOCK

It is always a good idea to approach studies into the thermal behaviour of residential buildings from various scales or dimensions of analysis. It is important to bear in mind that the success of building stock regeneration operations on an urban scale depends on the integrated coordination of several factors, as the European Economic and Social Committee [11] and the declaration of Toledo have already postulated. The situation is similar with respect to energy efficiency, as there are numerous factors involved in the inefficiency of Spain's housing stock. In this regard, we propose different scales of approach to the problem:

- The size or scale of the **user** and their energy habits, their education or degree of awareness regarding the management of resources, as well as social aspects that influence situations of vulnerability, such as an increased number of unemployed or elderly people, or concentrations of foreign populations with different habits.

In Spain, the ageing trend reflected by the population pyramid means we can predict an increase in vulnerable population groups in the coming years (Fig. **5**).

- The scale of the **home** as a first guarantee of comfort and habitability with which the user will directly interact. Several determining factors will affect its retrofit: whether it is owned or rented, if it is subject to charges or mortgaged, whether it is a first or second home, or if it is vacant or abandoned.

Presently, the demand for thermal comfort is not a priority for users when it comes to purchasing/selling or renting a home, although it is an essential parameter when it comes to determining the quality of life and a basic requirement that is not being met in some buildings. To date, it is impossible to find sufficiently clear information designed for the user, whether published by public authorities or by energy companies.

- The scale of the **building** whose construction properties determine the thermal behaviour of its envelope. Many improvement projects need to be implemented at the level of the entire building and, given the multi-family nature of the majority of residential buildings, the process needs to be coordinated between several owners.

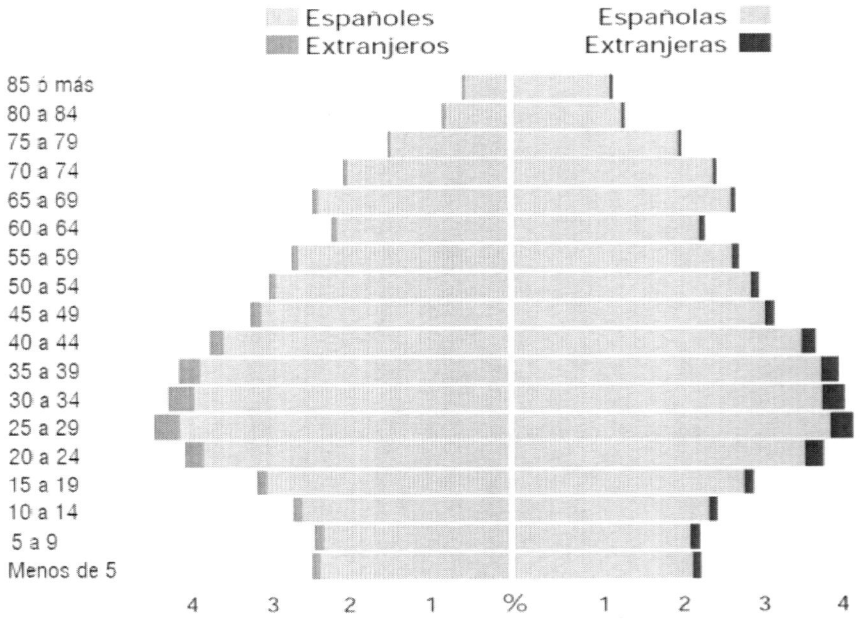

Figure 5: Spanish population pyramid in 2011. Source: *INE* [7].

While the size of the home and the building coincide in the case of single-family houses, the majority of homes constructed in Spain are not of this ilk. The city of Madrid accommodates a large percentage of the population in shared, high-rise buildings and social housing developments that were constructed during the study period (1940-80), both of which were multi-family developments. And so these types of residential buildings should be the focus of retrofitting projects that are designed to have a noticeable impact in terms of improved efficiency and reduced energy consumption.

- Finally, an in-depth analysis must be performed at the scale of the **neighbourhood**, in order to achieve a notable impact by improving energy efficiency and reducing consumption. It is therefore necessary to detect the areas which have the most inefficient building stock.

Furthermore, it is imperative that energy retrofitting is integrated into broader urban regeneration schemes so they may facilitate the horizontal resolution of other problems.

Household Energy Consumption

In Spain, a study by the IDAE [12] demonstrated that energy consumption is greater in single-family dwellings than in multi-family buildings or blocks. The study reported that, "The total consumption of a single-family dwelling is double that of homes in blocks, with four times more energy used for heating" (Fig. **6**).

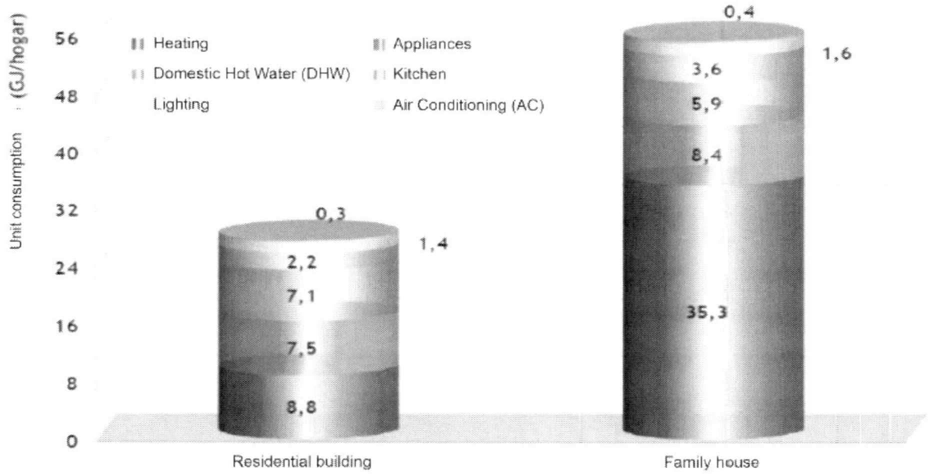

Figure 6: Energy consumption per unit according to the type of dwelling (IDAE 2011).

The decreased demand for energy used in heating and cooling also entails a positive impact on the requirements for generation and distribution infrastructures. The tendency for increasing levels of energy consumption in the residential sector, although attenuated in recent years by the economic crisis, implies the amplification of these infrastructures and, consequently, an increase in energy dependency and a heavier environmental impact.

According to the IDAE [13], the distribution of energy consumption based on usage in domestic buildings across Spain is as follows: 48.8% is used for temperature regulation, (heating 47% and air conditioning 1.1%), 27.4% for Domestic Hot Water (DHW), 20.6% for household appliances and 3.9% for lighting. This means that 50% of energy is destined to provide thermal comfort for the dwelling (see Fig. **7**).

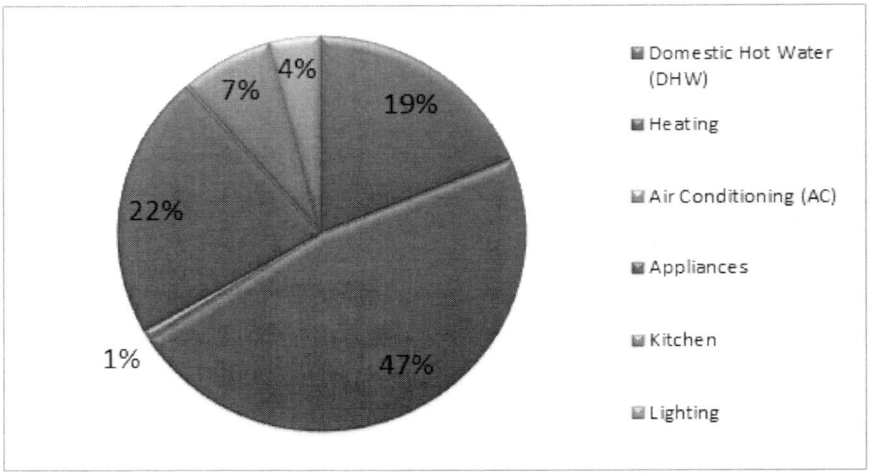

Figure 7: Distribution of energy consumption in Spanish households. (IDAE 2011).

The average Spanish household consumes 10,521 kWh/year, of which 65% is expended on thermal demands (heating, AC and cooking) and 35% on electricity [14].

It is important to remember that energy efficient retrofitting can represent a significant economical burden or upheaval for some sectors of the population. While the user also has a key role in terms of how the energy is used.

Energy Poverty

Although the average values presented in the previous section can serve to reveal the magnitude of energy consumption in Spanish homes, it is not always possible for a large number of households to meet the costs of maintaining comfortable living conditions and so their consumption is less. It is probably safe to assume that a significant part of the population who live in neighbourhoods constructed between 1940 and 1980 suffer from "energy poverty". In Spain, it has been estimated that 10% of homes found themselves in this situation in 2010 [15].

In the residential sector, energy poverty refers to the ability to maintain a warm home. The United Kingdom, which has dedicated more attention to this problem than any other European country, considers a household to be energy impoverished when it spends more than 10% of its income in maintaining the household at an adequate temperature. As a principal reference, the World Health Organization considers a comfortable temperature to be 21 °C in the living room and 18 °C in all other rooms.

In order to quantify and approach the problem, the European Commission proposed the following definition for energy poverty, while underlining that the definition could also incorporate other criteria in order to update the concept as society evolves: "*The difficulty or inability to maintain the home in suitable temperature conditions, as well as a lack of other essential energy services at a reasonable price*" [11].

In the context of Spain, the REPEX Project [15] combines the two concepts in the following definition: "Any homes that are incapable of paying for a sufficient volume of energy services to meet their domestic needs and/or that are forced to spend an excessive part of their income on energy bills for their homes (more than 10% of their annual income)". More recent studies situate this figure at 13% for the Autonomous Community of Madrid and associate the problem to the quality of the buildings [16] (Fig. **8**).

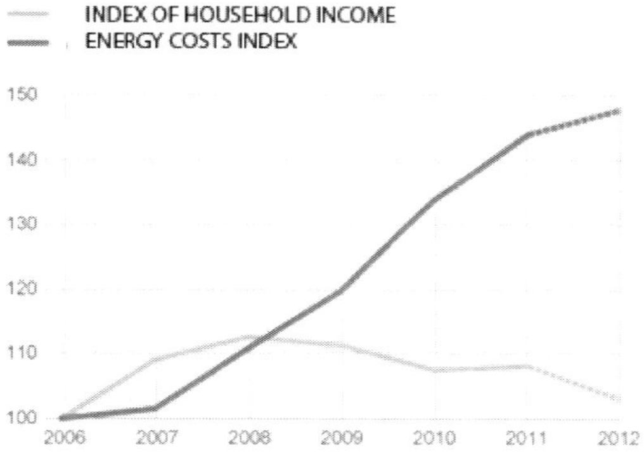

Figure 8: Household incomes and energy spending. (Source: Article in the newspaper *El País* (27/3/2014)).
http://sociedad.elpais.com/sociedad/2014/03/27/actualidad/1395947956_321445.html.

More recently [17], a more complex indicator has been developed for the United Kingdom: the LIHC (Low Income - High Cost), which defines energy poverty as the overlapping situations of low income and high energy costs. It considers energy poverty as a situation in which a household has energy costs above average energy requirements and an income that, after paying its energy bills, is below the poverty threshold (Fig. **9**).

The energy poverty suffered by these homes, therefore, mainly derives from the combination of three factors: a low level of income, poor quality buildings and high energy prices. In Spain's case, an additional factor is the overall lack of perception of this social problem, which impedes the drafting of any policies to improve the situation, although it is true that the civil society is starting to set up pressure groups and initiatives that provide visibility to the issue.

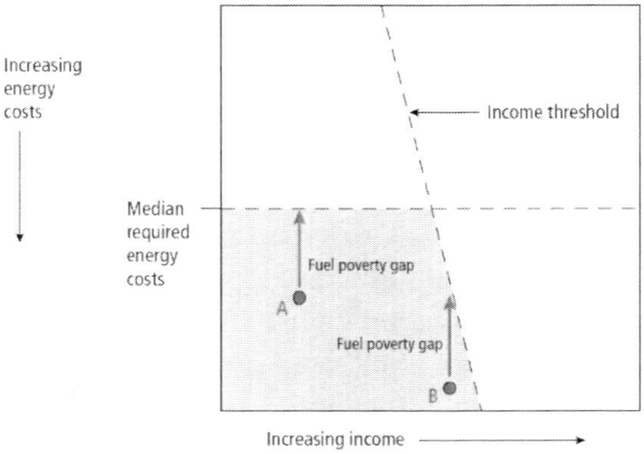

Figure 9: Calculation of the energy poverty gap. (Source: J. HILLS) [17]. Final Report of the Fuel Poverty Review.

The idea of energy poverty is commonly associated to the use of energy for heating, but the demands of other domestic needs must also be taken into account, such as maintaining an appropriate indoor temperature in summer (an important factor during the midday sun or heat waves when in hot climates), as well as requirements for lighting, household appliances, hot water, *etc*.

A situation of energy poverty can have multiple consequences, for example, the appearance of health problems [18] which can even lead to deaths in the most vulnerable groups, disconnections from the network by the supplier, rates of energy usage below the standards of comfort, or the accumulation of debt.

Solving the Energy Vulnerability of the Housing Stock

Energy poverty is a priority given that it has an influence on several current concerns, such as the alleviation of poverty, health, well-being, energy efficiency and a reduction in the emission of CO_2 and other pollutants. The European Commission has indicated that the measures destined to tackle this problem should follow a course that encourages "economic, social and territorial cohesion, as well as reasonable, transparent and comparable prices among the different suppliers". The renovation of vulnerable housing, which includes an element of energy efficiency improvement, means a long-term solution to this problem can be proposed.

A building or dwelling is considered to be energy vulnerable when it is inefficient and requires large amounts of energy to achieve normal standards of comfort. In this sense, energy demand is a good indicator for characterising a dwelling's vulnerability. When determining the basic requirements for interior thermal comfort under standard conditions, energy demand is an indicator that depends on the building's architectural characteristics and urban environment [19]. A vulnerable dwelling is understood to be one which behaves inefficiently because it has high energy demands.

The policies aimed at energy efficient housing have the potential to: reduce energy poverty; significantly reduce consumption; and minimise vulnerability to unforeseen changes in household income and/or to unstable and growing energy prices. The reduction of energy poverty adds a key point to the environmental and economic objectives for energy efficiency improvements in residential buildings.

The existing building stock gives rise to high levels of energy consumption as a result of its inefficiency due to poor thermal quality in the majority of the constructions and due to the shortcomings of their HVAC (heating, ventilation and air conditioning) systems. As mentioned earlier, a 2012 summary of basic information on consumption in the residential sector prepared by the IDAE [12] situates average annual energy expenditure per household at €990, of which 47% corresponds to heating, thus it is easy to deduce that households in a situation of poverty will have lower expenditure. The most vulnerable social groups coincide with those of least income, *e.g.*, people over 65, single-parent families, unemployed people or those who collect social benefits. There is also a correlation, in the majority of cases, between people with low levels of income and those living in buildings with inadequate thermal insulation [20], hence the situation of energy poverty is exacerbated.

As stated previously, in Spain's case the problem derives from a building stock constructed with poor quality materials in terms of energy efficiency, these dwellings were constructed before the introduction of standard NBE-CT-79 or even before the Royal Decree on Energy Saving 1490/1975, which didn't consider thermal insulation. The largest volume of housing was built in the years from 1961 to 1980, a period of prolific construction of high-rise blocks that accommodated a larger number of smaller dwellings compared to previous periods.

Some studies also indicate that the thermal quality cannot necessarily be guaranteed in the case of buildings erected in the 90s construction boom [21]. The majority were constructed before the Technical Building Code came into force, which has since raised energy efficiency standards.

> *"Although in principle one could believe that by complying with Basic Standards for Buildings, Thermal Conditions (NBE-CT-79) a building's thermal characteristics would be improved, but this is not strictly correct since the speed with which a huge number of buildings were constructed, the low-quality materials or the lack of qualified workers means that these buildings also have significant energy deficiencies"* [21].

Attempts were made to compensate for the problem of poor thermal quality in Spanish building by trusting comfort to the installation of unnecessary and oversized equipment, regardless of the high energy consumption levels and increases in greenhouse gas emissions. Systems with low performance, and, therefore, high levels of consumption and CO_2 emissions, are considered inadequate.

> *"With the widespread introduction of electro-mechanical AC systems in the 1970s, efforts were made to definitively reduce the weight of the facades, almost exclusively installing an exterior skin measuring ½ a foot thick, giving rise to the situation we now face today"* [22].

Heavy construction was abandoned during the 20th century and so the thickness of facades reduced dramatically. Thermal comfort became a problem that was resolved through the use of HVAC systems. Low energy prices and confidence in unlimited resources guaranteed several years of wastefulness until the 1970s oil crisis, a point of inflection which led to Spain introducing energy-saving policies and the obligatory use of insulation systems in building projects.

This problem was not resolved when the basic standard came into force in 1979 because from that date forward developments in HVAC systems eclipsed the need to improve the passive thermal characteristics of new constructions.

> *"The 90s construction boom coincided with a period of peak development in HVAC technologies, leading to excessive installation of unnecessary equipment in an attempt to compensate for the shortcomings of low-quality constructions. As such, many of these buildings contain equipment and installations that are oversized, inefficient and require high rates of consumption to heat/cool the dwellings."* [21]

Summertime energy expenditure has rocketed due to the popularisation and indiscriminate use of air conditioning systems.

Coupled to these inefficiencies in buildings and their HVAC systems, Spain is living through an endemic problem that is also observed in the hotter Mediterranean zone. Comfortable temperatures are easily achieved throughout the majority of the year, but as Tirado *et al.* explain:

> *"In these areas of generally more benign climate, which only suffer a few weeks of cold weather each year, the dwellings are poorly insulated and often lack an adequate heating system, above all in the case of low-quality housing occupied by low-income families. These households have to use electric heaters (which produce heat at a much higher cost per unit) or other suboptimal forms of heating that do not warm the living space sufficiently, but which still imply a significant burden on family budgets during this cold period. Although the annual weight of energy bills in the domestic budget is relatively low, many households would have endured a state of energy poverty during the weeks in which heating was necessary. These households are exactly those which, in the regions with a warmer climate, would be categorised as incapable of maintaining the home at an adequate temperature in accordance with the ECV"* [15].

The REPEX project aims to shed light on the problems faced by these climates, classifying energy poverty into two types according to the nature of the analysed indicators. Their study draws relations between the subjective indicators based on the level of the occupants' thermal discomfort and the lack of adequate heating systems.

CONSTRUCTION OF FACADE ELEMENTS

In residential buildings constructed in the municipality of Madrid during the study period, the envelope element with the greatest surface area is the facade, separating the interior and exterior environments by means of opaque components and glazed openings.

As seen in Fig. **10**, many of these constructions were high-rise buildings. Only 3.5% of the dwellings were constructed in single or two-storey buildings, 5% were three-storey buildings and 13% buildings were constructed over four floors. Five-storey residential buildings account for 25% of those constructed throughout the study period, with a large number built during the 1960s. Over half of the building stock, 54%, has six or more floors, with these higher buildings representing the most numerous types to be built in the last two decades.

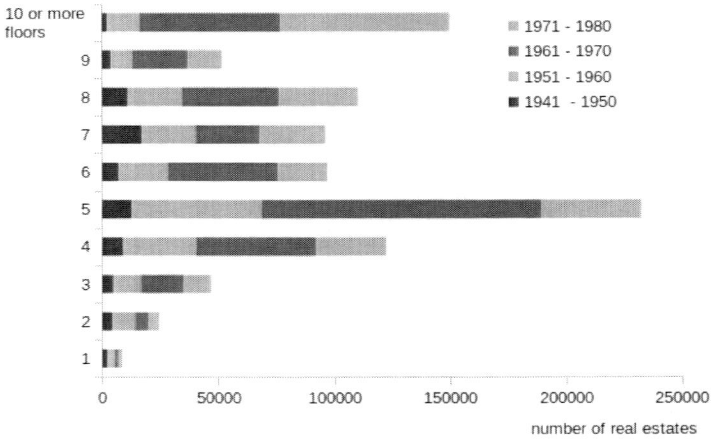

Figure 10: For the municipality of Madrid, number of households in primarily residential buildings according to the building's number of floors and decade of construction. (Data from the 2011 census).

A literature review and some on-site inspections have been carried out in order to appreciate which construction solutions are most commonly used, to understand their energy performance and to propose suggestions for retrofitting solutions.

Using the original technical documentation from each case, we have reviewed 53 projects with the following temporal distribution: from 1940 to 1950, 9 projects; 1950-60, 26 projects; 1960-70, 13 projects; and from 1970 to 1980 there were 5 projects. We deduced the composition of the blind walls of the facades, as well as the types of windows and their protections, *i.e.*, the type of frames, the type of glass and whether or not they had blinds. Data are grouped into the four decades involved in the study, allowing us to draw independent conclusions about the construction typologies employed in each decade, as well as information about the overall typology used across the 40 year period [23].

From the initial 53 projects, we selected a series of buildings representative of the most used types of facade and then physically inspected them to determine their state of conservation and whether their facades were suffering any pathological deterioration processes.

Opaque Components

Regarding the types of construction used in the facades we inspected, in accordance with their original project descriptions, there were numerous variations which can be categorised into four basic groups:

- single-skin brick facades of varying thicknesses, representing a total of 23 projects,
- double-skin brick facades separated by an air cavity, 24 projects,
- double-skin brick facades with intermediate insulation, 4 projects, and
- prefabricated concrete facades, just one project.

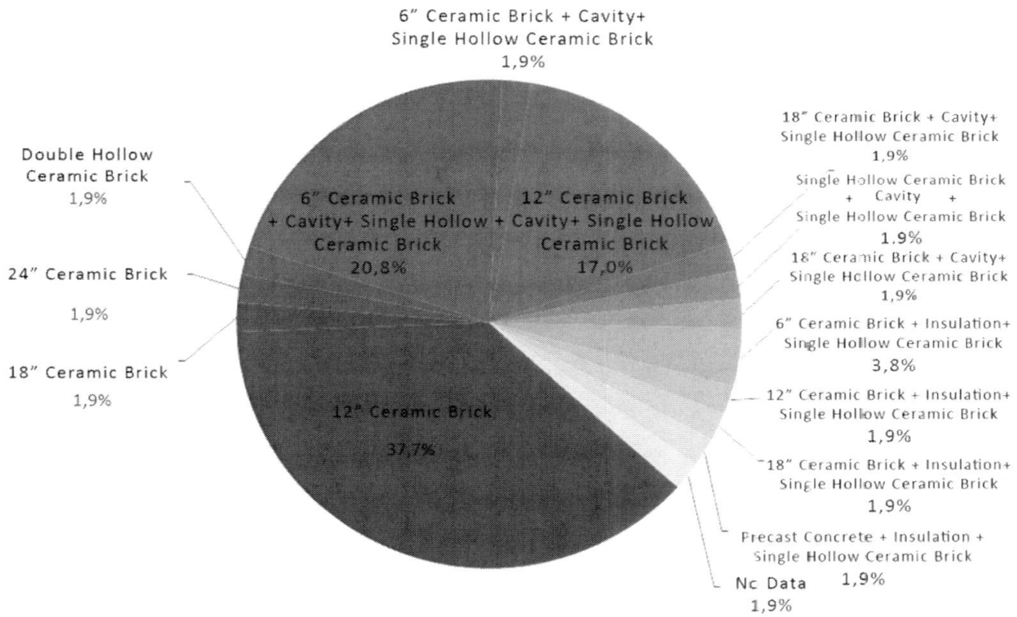

Figure 11: Distribution of the different types of opaque facade in a sample taken from Madrid during the 1940-80 study period. (Compiled by the author).

The blind enclosing walls of facades in buildings constructed from 1940 to 1980 as part of publicly protected housing schemes (Fig. **11)** have developed from a single solid skin of 1-foot of brick into a double-skin envelope, firstly with an air cavity and then later, in a few cases, with intermediate insulation. In double-skin systems, the exterior wall was reduced from 1 foot to ½ foot, while during the study period the interior wall continued to be constructed with SHCB. This became the typical solution for residential buildings, even for private and unsubsidised buildings, until the 1980s when the interior skin was replaced by DHCB.

Glazed Openings

We analysed the frames, the type of glass and whether or not any solar protection was used in the glazed openings of the building envelopes involved in the projects evaluated.

The frames used in these types of buildings and in this period were mainly wooden, although there are plenty of examples of metal frames. Single pane windows between 3 and 5 mm thick were used. A high percentage (40%) have external blinds offering protection from the sun. In general these windows provide little in the way of insulation, so many have been replaced over the years, especially in order to change the type of glass and improve the energy behaviour. Similarly, blinds have been subjected to conservation and replacement, or incorporated in cases where they were absent (Fig. **12**).

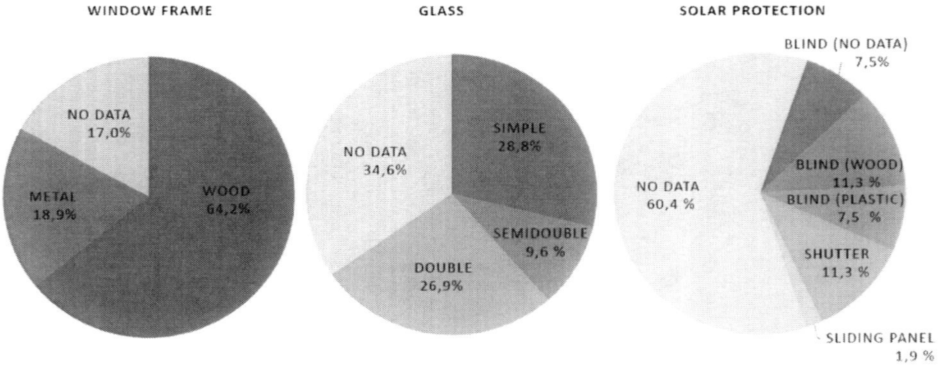

Figure 12: Distribution of different types of glazed openings in a sample taken from Madrid during the 1940-80 study period (Compiled by the author).

THERMAL CHARACTERISATION OF OPAQUE COMPONENTS

Retrofitting can be contemplated in an integral manner or according to individual elements. Although both methods have advantages and disadvantages, the present study attempts to isolate the retrofitting objective to the scale of construction elements [24, 25].

We can distinguish the key factors which shall determine the thermal behaviour: the climate, the urban environment and the user's habits. The building's response to these conditions will depend on its design, its HVAC systems and the constructive composition of its envelope; this response will subsequently determine the internal degree of comfort and well-being. A detailed study of the thermal behaviour offers a starting point from which to propose improvements.

Although openings and glazed areas exchange the largest amounts of energy (per unit area) *via* their connection with the exterior (sun, air, light), it is opaque sections that have the greatest surface area and which tend to be the most difficult for the user to approach for retrofitting purposes. It is also, therefore, the element which requires the most durable solutions.

Transmittance of Opaque Enclosures: Theoretical Calculation

From the review of 53 housing projects in Madrid, the thermal transmittance U (W/m^2K) calculated for a stationary system according to standard UNE-EN ISO 6946 and the Spanish Technical Building Code for opaque enclosures is presented below.

As can be seen in Fig. **13**, the group with the highest thermal transmittance consists of buildings with facades constructed from a single layer of solid bricks. Those with two layers separated by an air cavity yield better performance. Relevant data can be extracted by examining the types of facades organised according to construction year. A tendency for increasing degrees of insulation is apparent, with more cases of double-skin facades built from the 50s onwards. Some envelopes that included a thin layer of insulating material can also be found towards the end of the same decade.

Within this group of cases, there wasn't a single instance of one layer, 15 cm thick facades constructed from either solid or perforated brick, which would have been thermally poor buildings and also more susceptible to humidity through the effects of condensation.

The two cases of single-layer enclosures thicker than 30 cm were presumably constructed with perforated brick, which is why their thermal resistance is similar to that of a double-skin facade. If they were constructed with solid bricks they would conduct more heat. The double-skin facades have thermal transmittances of 1–2 W/m^2K and the most resistant are, logically, those with some degree of insulation, which can range from a sheet of cork to 40 mm of insulation.

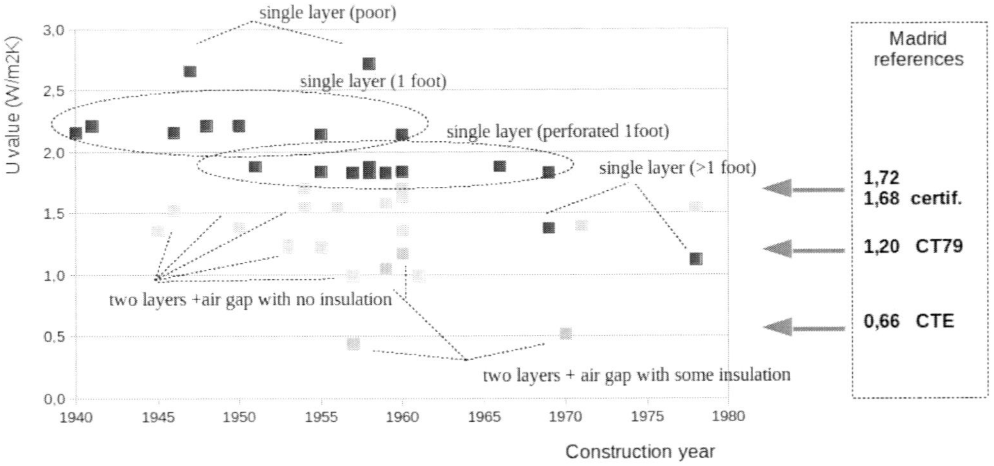

Figure 13: Thermal transmittance U (W/m^2K) according to the construction year for the cases studied and references for Madrid. (Compiled by the author).

The current reference for maximum transmittance in Madrid is set by the CTE at a value of $0.66 W/m^2K$. Practically all the facades studied were above this level, the majority were even above the standard given for Madrid by NBE-CT-70: 1.20 W/m^2K for light facades and $1.40 W/m^2K$ for heavy facades (< 200 kg/m^2).

Values currently used as references for energy certification in existing buildings are as follows: for buildings constructed in the period 1940-60, 30 cm of 260 mm perforated bricks with 20 mm of plaster, with a transmittance of 1.68 W/m^2K; and for buildings from 1960 to 1979, a double-skin facade with a 15 cm of 120 mm perforated bricks, a 30 mm air cavity and a layer of 50 mm single hollow ceramic brick with 20 mm of plaster, with an overall transmittance of $1.72 W/m^2K$. These references represent the best cases observed in this study for single-layer enclosures and the worst cases for double-skin facades. It is important to remember that the thermal resistance of perforated bricks is significantly greater than that of solid bricks and in many cases the thickness is specified but the type of brick isn't; hence, there may be variations.

Thermal Inertia. Simulation of the Gap and Thermal Attenuation

Thermal mass is one of the factors involved in the internal thermal stability of housing and, along with temperature variations, defines the amount of thermal energy that can be stored by the building envelope. This factor depends on the volume, density and specific heat of the construction materials [26]. Energy accumulation within the building envelope, or the thermal inertia, produces a displacement and attenuation with respect to the external thermal wave, and, therefore, is directly related to internal temperature stability.

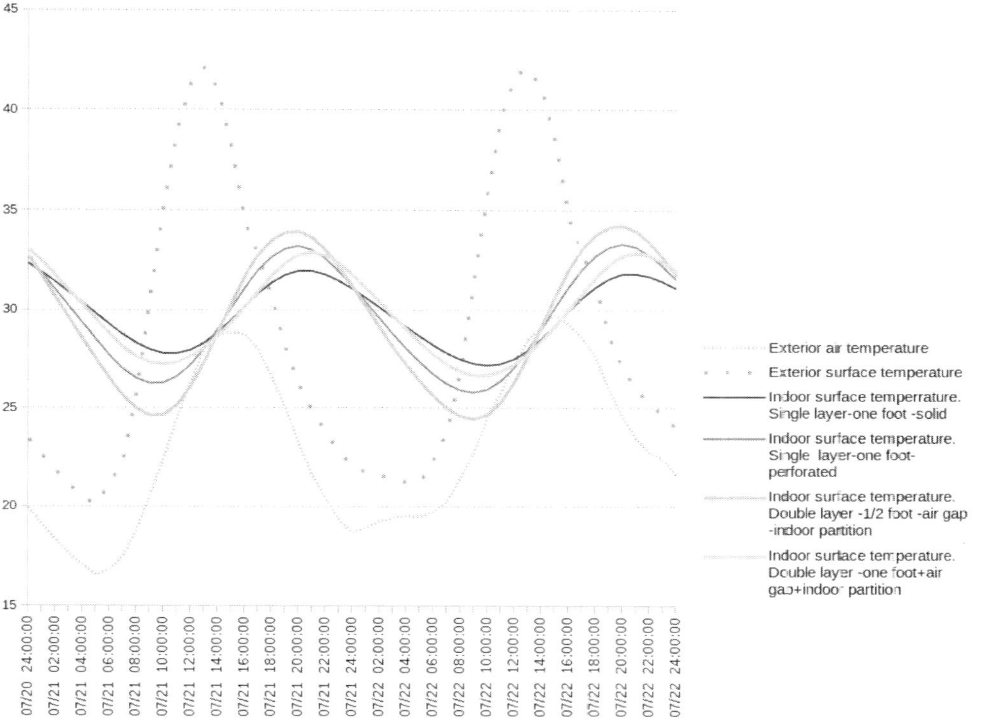

Figure 14: Simulation of the daily thermal wave for different types of facade over two summer days. Facade surface temperatures and ambient temperature (°C). (Complied by the author).

Fig. **14** shows the evolution of the internal surface temperature over the course of 2 summer days for 4 different facades experiencing the same conditions and each facing south. It deals with the commonest types of facade in the sample: single layer, 30 cm of perforated brick; single layer, 30 cm of solid brick; double-skin, 30 cm of perforated brick with a cavity and a layer of single hollow ceramic brick; and double-skin, 15 cm of perforated brick with a cavity and a layer of single hollow ceramic brick. Although several factors play a part in the simulation, so that we may observe the effect of the thermal inertia, the dynamic behaviour is presented for a summer's day and so one of the most important factors is solar radiation. Data for the external surface temperature have also been included in order to demonstrate the temperature increase caused by incident radiation falling on the facade. This effect is clearly greatest around midday.

Thermal attenuation differs for each case in the study. The heaviest facade, composed of 30 cm of solid brick, provides the highest thermal stability, as shown in Fig. **14**. Facades built with 30 cm of perforated brick also have an attenuating effect, only slightly less than the aforementioned facade, but still significant. If a layer of single hollow ceramic brick plus an air cavity is added to this 30 cm layer of perforated brick, then the displacement of the wave can be observed. In other words, the facade takes longer to heat up and cool down, due to a greater thermal inertia. The lightest facade, in the present study represented by the double-skin facade with 15 cm of external wall, presents the greatest amplitude of oscillation and stores the least amount of heat.

This effect of thermal inertia is of particular interest in winter months when the daily oscillation and the combination of passive strategies can provide a greater number of hours of comfort without resorting to active heating systems, thus reducing usage of and dependence on these systems. When it comes to proposing energy retrofitting strategies, the insulation's characteristics and positioning will also take on enormous importance.

The various types of facade encountered indicates the importance of and need for a detailed study into each particular case when characterising the thermal behaviour and proposing retrofitting strategies. Such detailed studies will allow us to characterise the solutions' suitability and efficacy. Single-skin facades with a reduced thermal resistance are the oldest in our study and will thus result in increased heating and cooling demands. The facades with greater thermal inertia are heavier and, therefore, will provide a better response to temperature fluctuations and radiation effects, helping to reduce thermal loads, especially in summer.

Possibility of Condensation: Theoretical Calculation

This calculation has been made as per the provisions set out in current regulations. The possibility of surface condensation, according to the CTE, depends on the internal relative humidity and the average external temperature, calculated for the month of January. In the case of Madrid, which in January has an average external temperature of 6.2°C and an internal relative humidity of 55%, surface condensation will occur on facades with a transmittance U > 1.71 W/m^{2}K. This limit separates the majority of the single-skin facades from the double-skin envelopes. The transmittance limit is lower in cooler regions (*e.g.*, in León U> 1.40 W/m^{2}K) compared to warmer ones (*e.g.*, in Seville U > 2.54 W/m^{2}K), which has presumably resulted in differences between the construction, maintenance and restoration of this type of facades. An increase in the internal relative humidity caused, for example, by a certain type of use or the degree of ventilation, also affects the transmittance limit.

The possibility of interstitial condensation has been calculated for the different types of double-skin facades with no insulation. According to calculation data from the CTE for Madrid, there is no risk of interstitial condensation on these types of facade. A temperature and pressure profile for each layer in such a case is shown in Fig. **15**.

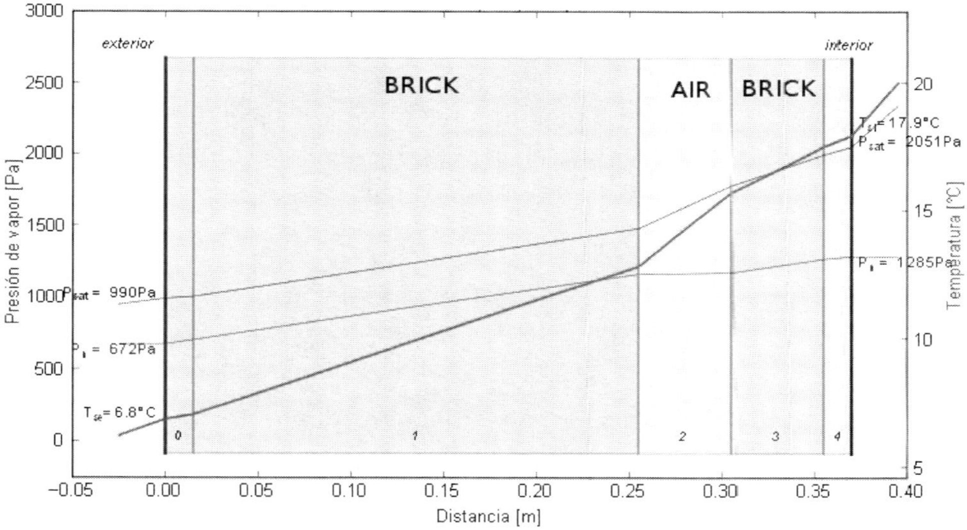

Figure 15: Temperature, vapour pressure and saturation vapour pressure distributions for a double-skin facade. (Compiled by the author). (Software condensaciones v0.60).

This type of profile reflects a certain climate and does not present any jumps in the pressure profiles, unlike those which occur when using insulation or barriers that are less permeable to water vapour. The use of insulation, either internally or in the cavity, changes the pressure profiles and also the possibility of interstitial condensation.

In general, the possibility of surface condensation is greater in single-skin facades than in double-skin facades, because the former typically provide less insulation.

The absence of any barriers that would increase the vapour pressure in the facade's interfaces prevents the possibility of condensation, an important point to bear in mind when selecting energy retrofitting strategies.

PROPOSALS FOR RETROFITTING SOCIAL HOUSING

The data from 53 housing developments constructed between 1940 and 1980 have been collected. Among the studied cases, the group with the lowest thermal resistance consists of buildings with facades constructed from a single layer of solid bricks, followed by those with a single layer of perforated bricks. The most resistant are, logically, double-skin facades with some degree of insulation. The two most abundant construction solutions are that of a single layer with 30 cm of brick and the double-skin facade with a 15 cm exterior brick wall, an air cavity and a lining of SHCB.

In view of the construction typology encountered, and of its deficient energy behaviour in the majority of cases, it makes sense to consider improving the facades of practically all the buildings that have not yet received some form of overall renovation. Their thermal resistance is below the level stipulated in current regulations and standards, which are increasingly rigorous.

The Priority for Passive Measures

Due to this inefficient performance of the opaque vertical facades, coupled with the lack of general maintenance in social residential buildings and the deterioration of the neighbourhoods in which they are found, the most appropriate improvement measures are directed towards retrofitting the opaque building envelope at a neighbourhood-wide scale, including elements to improve thermal behaviour and reduce energy demand.

Facade retrofits should be approached by considering the building as a whole unit. The priority must lie in adapting the building's passive systems, and thus increasing thermal comfort inside the dwellings. After reducing energy needs, attempts should then be made to improve the installations and energy-consuming equipment, as they would now have lower demands to satisfy. Having improved these aspects, the next step would be the incorporation of energy systems that rely on renewable sources.

Retrofitting by incorporating exterior thermal insulation will not only improve the building's thermal resistance but also its properties insofar as the accumulation of heat and the risk of condensation, while also eliminating any thermal bridges on structural interfaces. The placement and characteristics of the solutions proposed for the retrofit, therefore, must be adapted to each type of enclosure.

A Life-Cycle Analysis for the Retrofit

When considering the options for energy retrofits, it is also important to incorporate a life-cycle analysis of the dwellings and their construction elements. This will complete the energy evaluation and environmental impact analysis for the retrofit.

As observed in other studies [27-29], the energy incorporated within the construction materials used in these facades (such as glass, ceramic, mortar, or various metals) can be very relevant to the overall picture, in terms of both evaluating the existing dwelling as a resource and the environmental impact of possible retrofitting solutions. Within the life-cycle analysis, it would be particularly interesting to characterise the maintenance and durability of the facade solutions.

The Need to Inform and Public Participation

It is evident that a sufficiently broad group of actors need to participate in the decision-making process for it to be representative of the lattice of diverse interests, influences and intentions surrounding the question of energy retrofitting.

Public administrations are making an effort to apply the European energy efficiency directives, there is hope in the creation of "green employment", and businesses are showing an interest in the possibility of opening new markets, but the final users, who will be expected to pay for these improvements, are receiving rather limited information. The Building Report, the *ITE* (Technical Building Inspection), the Energy Certificate, and the Building Record, all have a certain obligation to inform and educate the user about the correct use and maintenance of the construction. However in reality, efforts to educate the public have been insufficient, and these tools are usually seen as a charge or tax rather than a tool to help us extract maximum benefit from our energy resources and our buildings.

When addressing energy retrofits, householders must be able to count on reliable and independent decision-making tools and information. The aim is to demonstrate to the owners of the buildings that investing in energy efficient retrofitting, apart from reducing greenhouse gas emissions in accordance with international agreements, will also have direct positive consequences on their quality of life.

The Responsibility of the State

Universal access to energy is a responsibility of the public administrations from the moment it is perceived as a necessity without which survival would be threatened. The growing number of studies into how a lack of comfort affects human health, or those about climate-associated seasonal mortality, demonstrate that energy is a basic need along with food, clothing and decent housing.

This state responsibility is particularly evident in social housing, where they must cover the basic efficiency, comfort and resource-saving standards as required by current regulations. To achieve this it is necessary to allocate public resources in order to retrofit the inefficient public housing stock, especially in those cases where the inhabitants struggle to pay their energy bills and, therefore, are not prepared to foot the bill for energy retrofitting.

ACKNOWLEDGEMENTS

The present study has been carried out at the IETcc-CSIC and forms part of project BIA-2012-39020-C02-01 funded by the Spanish Ministry of Economy and Competitiveness. Data from this work form part of Project No. C33/06 Ministry of Internal Development, FOM/2251/2006. This work has been developed using figures from different presentations given by the authors at national and international Congresses and Conferences, see references.

CONFLICT OF INTEREST

The authors confirm that this chapter contents have no conflict of interest.

REFERENCES

[1] RD.2429. NBE-CT-79, "*Real Decreto 2429/1979, de 6 de julio, por el que se aprueba la norma básica de edificación NBE-CT-79, sobre condiciones térmicas en los edificios*". Spain: Presidencia del Gobierno, 1979.

[2] L. Moya, "*Barrios de Promoción Oficial. Madrid 1939-1976*". Madrid: Colegio Oficial de Arquitectos de Madrid, 1983.

[3] C. Sambricio. "*Un siglo de vivienda social: 1903-2003*" in [exposición organizada por el Ministerio de Fomento, Ayuntamiento de Madrid y el Consejo Económico y Social (CES)] Sambricio, C. [comisario y editor] Madrid, Spain, 2003.

[4] P. Cassinello, Eduardo Torroja 1949. "*Strategy to industrialise housing in post-World War II*". Madrid: Fundación Eduardo Torroja, 2013.

[5] J.A. Domínguez, V. Azorín, "*La vivienda social después de la guerra civil. El Inicio de la Modernidad*". Madrid: IETcc-CSIC, Unpublished work, 2006.

[6] European Parliament and the Council of the European Union. "*Directive 2010/31/EU on the energy performance of buildings (EPBD recast)*". Official Journal of the European Union. 2010.

[7] Spanish National Institute of Statistics (INE). 2011. "*Census on population and housing*". http://www.ine.es/censos2011_datos/cen11_datos_inicio.htm [Accessed Feb 15, 2015].

[8] I. Oteiza, C. Alonso, "*Hacia una rehabilitación de la vivienda social en las grandes ciudades españolas, a través de la eficiencia energética de la envolvente. Madrid 1940-1980*" in I Congreso Internacional y III Nacional de Construcción Sostenible y Soluciones eco-eficientes, Seville, 2013:

[9] "*RD.314. REAL DECRETO 314/2006, de 17 de marzo, por el que se aprueba el Código Técnico de la Edificación. Ministerio de vivienda*". Spain, 2006.

[10] A. Cuchi and P. Sweatman, "*Una visión-pais para el sector de la edificación en España. Plan de acción para un nuevo sector de la vivienda*". Madrid: Green Building Council Spain and the Conama Foundation. 2012. Rehabilitation Working Group Report 2012. ISBN 978-84-616-1917-7, 2012.

[11] European Economic and Social Committee. Opinion of the European Economic and Social Committee on "*The need to apply an integrated approach to urban regeneration*" (exploratory opinion). 21.01.2011. Official Journal of the European Union C 21/01. 2011

[12] IDAE. "*Análisis del consumo energético del sector residencial en España*". SECH-SPAHOUSEC project. 2011< http://www.idae.es/> (Accessed Jan. 2013).

[13] IDAE. "*Plan de acción de ahorro y eficiencia energética 2011-2020*" < http://www.idae.es/> 2011 (Accessed Feb. 2015).

[14] R. Fernández, "*EnergyLab- Presentación*" in Jornada Eficiencia ambiental. Gas natural-Fenosa. 7/11/2012. Data provided by the IDAE and the INE, 2012.

[15] S. Tirado Herrero, S. López Fernandez and P. Martín García, "*Pobreza energética en España. Potencial de generación de empleo derivado de la rehabilitación energética de viviendas*". Madrid: Asociación de Ciencias Ambientales, ACA, 2012.

[16] C. Sánchez-Guevara, A. Sanz Fernández, A. Hernández Aja, "*Income, energy expenditure and housing in Madrid: retrofitting policy implications*", Building Research & Information, 2014. DOI: 10.1080/09613218.2014.984573.

[17] J. Hills, "*Getting the measure of fuel poverty. Centre for Analysis of Social Exclusion*", Final Report of the Fuel Poverty Review, London: The London School of Economics and Political Science. Department of Energy and Climate Change (DECC), p. 237, 2012. ISSN 1465-3001.

[18] J. D. Healy, "Excess winter mortality in Europe: a cross country analysis identifying key risk factors s.l.: BMJ Group", *Journal of Epidemiology & Community Health*, vol. 57, pp. 784–789, 2003.

[19] F. Martin-Consuegra, "*Vulnerabilidad energética asociada a la edificación. Estudio de caso para la periferia urbana del sur de Madrid, Territorios en Formación N06*", Madrid, Spain, pp. 105-118, 2014.

[20] O. Lelkes and E. Zólyomi, "*Housing Quality Deficiencies and the Link to Income in the EU. Policy Briefs*". Viena: European Centre for Social Welfare Policy and Research, pp. 20, March 2010.

[21] M. Dalle, G. A. Portillo, C. A. Palacios, L. B. García, B. E. López, A. M. Marco, B. M. García, M. M. M. García, E. M. Bértolo. "*La generación de empleo en la rehabilitación y modernización energética de edificios y viviendas*". Madrid: EUROPEAN UNION, European Social Fund. Fundación Biodiversidad. s.l.: Fundación Conde del Valle de Salazar, pp. 198 V-0000-2011, 2010

[22] J. Monjo Carrió, "La evolución de los sistemas constructivos en la edificación. Procedimientos para su industrialización", *Informes de la Construcción*, vol. 57(499-500), pp. 37-54, Madrid 2005. doi:10.3989/ic.2005.v57.i499-500.481.

[23] M.A. Gonzalez Moya, C. Alonso, J. Monjo, I. Oteiza, "*Tipología constructiva de fachadas de viviendas de protección pública en Madrid. Décadas de 1940-1980*" Madrid: Eduardo Torroja Institute of Construction Sciences. Spanish National Research Council in Jornadas internacionales de investigación en construcción. Vivienda, pasado, presente y futuro, 2013.

[24] C. Alonso, I. Oteiza, "*Energy efficiency in Spanish social housing stock. Review of façade composition and energy demand, recent literature BESS-SB13*". Building Enclosure Sustainability Symposium. California: California State Polytechnic University, Pomona in Sustainable Buildings Conference, 2013.

[25] C. Alonso, M. González, I. Oteiza, J. Monjo, "*Hacia una rehabilitación de la vivienda social; transmitancia e inercia térmica de la envolvente vertical de edificios singulares en Madrid 1939-1979. Rehabend*". Santander: University of Cantabria in Latin American congress of pathologies in construction, retrofitting technologies and heritage management, 2014.

[26] F.J. Neila, *Arquitectura bioclimática en un entorno sostenible*. Madrid: Munilla Lería, 2004.

[27] C. Alonso, I. Oteiza, J. García, "Environmental analysis of residential building facades through energy consumption, GHG emissions and costs". *In Proceedings of SB11 Helsinki World Sustainable Building Conference*, 2011.

[28] C. Alonso, I. Oteiza, J. García, "*Diseño eficiente de elementos constructivos: Fachadas*". Seville: University of Seville in I Jornada Nacional de construcción sostenible y soluciones ecoeficientes, 2011.

[29] C. Alonso, I. Oteiza, J. García, "Design strategies in facades for the reduction of housing energy consumption: *In Proceedings of SB10*". Finland: Sustainable community - Building smart, 2010.

CHAPTER 2

Environmental Impact Indicator System in Residential Neighbourhood Renovation

Pilar Mercader-Moyano[1,*], Julia Garrido-Piñero[1] and Antonio Ramírez de Arellano[2]

[1]University of Seville, Department of Building Constructions I, Reina Mercedes 2 Avenue, 41012, Seville, Spain and [2]University of Seville, Department of Building Constructions II, Reina Mercedes 4 Avenue, 41012, Seville, Spain

Abstract: There is a specific urban fabric within Spanish towns that consists of neighbourhoods, built in the twentieth century between the late fifties to the mid-seventies, made of open blocks. These neighbourhoods (or 'barriadas') have inherited the state's answer to the major housing shortage at the end of the wars of the mid-twentieth century: work processes with short execution times, reduced costs, poor and emerging industrialised production, and comfort standards greatly inferior to today's minimum demands. These factors may lead to the removal of these occupying households from the housing market, or to condemn a large number of dwellings to marginality. Within these neighbourhoods, this research focuses on those whose design is the result of the repetition of one or more residential typologies of collective housing. Thus, residential complexes of urban character are set up and can be observed as episodes within the city rather than as autonomous elements. The ultimate goal of this research is the assessment and minimisation of environmental impact caused by these types of fabrics in cities. The objective is to identify the key areas that contribute towards achieving environmental refurbishment. To this end, a system is proposed of indicators that focus on those areas responsible for the imbalances produced in urban ecosystems. This indicator system is based on a comprehensive study of the current systems of both urban and building indicators of sustainability. As a result, a new system of indicators adapted to this specific situation is developed.

Keywords: Atmosphere, energy cycle, energy efficiency, energy savings, environmental assessment, environmental impact, environmental rehabilitation, indicator classification, indicator system, legislation, local characteristics, neighbourhood renovation, residential neighbourhood, residential typologies, strategic areas, urban design, urban fabrics, urban sustainability, waste cycle, water cycle.

***Corresponding author Pilar Mercader-Moyano:** University of Seville, Department of Building Construction I, Reina Mercedes 2 Avenue, 41012, Seville, Spain; Email: pmm@us.es

Pilar Mercader-Moyano (Ed.)

INTRODUCTION

Among the types of Spanish buildings, the collective building typologies are considered the most significant in relation to achieving the objectives of ensuring the quality and sustainability of the construction itself, provided that its use is residential or assimilated [1]. In fact, a high percentage of households in Spain live in apartment blocks [2]. Within this classification of buildings, a specific fabric can be found: the residential neighbourhoods (or "barriadas"). This phenomenon refers to social housing estates built by the Franco regime between 1957 (the year of the creation of The Department of Housing) and 1975 [3]. These neighbourhoods have inherited a reality which is the consequence of the housing needs at the end of the mid-twentieth-century wars. Furthermore, these housing estates are found in all kinds of cities in which there have been public interventions to solve housing problems. However, these interventions are significantly more common in medium-sized and large cities that have experienced a higher growth in the decades of the fifties and sixties. These neighbourhoods now represent a significant part of the problematic neighbourhoods, as can be observed in other industrialised countries [4].

The residential neighbourhoods define an urban area constituted by the buildings and their environment, and they create spaces in which people live and work, providing a setting that serves as an identity of the inhabitants. On the other hand, these urban areas also imply a huge consumption of natural resources, water, and other raw materials. For this reason, the role played by buildings and cities is considered fundamental in achieving sustainable development.

Substantially, 40% of CO_2 emissions in Spain are produced by urban areas; within this 40% are emissions produced by energy consumption in the domestic sector and building-related activities (work, commerce, leisure, *etc.*). Moreover, the building industry as a whole represents 42% of European energy consumption, with an average increase of 1.5%. Buildings represent, in fact, a major part of the city's energy consumption and, by extension, of society [2]. The challenge therefore is to minimise energy consumption and its associated environmental impact.

The Use of Indicators for Sustainability Assessment and Environmental Impact Assessment: Pros and Cons

Indicator systems allow us to understand complicated realities in a systematic way, since the indicators are estimations or statistics which help us to interpret complex data. Among their advantages, it is observed that these systems ease the development of the concept of sustainability in architecture, and serve as a powerful tool for the generation of awareness and knowledge. This is due to the fact that indicator systems simplify complex realities, and hence the information reaches a larger public. The indicator systems are analysis tools based on observable and quantifiable circumstances that allow objective quantification, and are backed by numerous verified experiences.

However, indicator systems also have disadvantages. Among others, there lies the extreme complexity of the issues discussed, which occasionally renders it difficult to simplify concepts. Therefore, it could be difficult to identify certain aspects of the reality objectively, and there is also the induced possibility of forced identification of the indicator as a rule, and furthermore the increased responsibility for the actors in the construction can be considered as another drawback of indicator systems [5].

Study Objective

The main aim of the research is to define an indicator system that allows us to evaluate and minimise the environmental impact in residential neighbourhood renovations, whereby this system is based on the most relevant strategic areas achieved by the study and on the analysis of existing systems. The last goal involves the ability to be able to reply the following question: **Which are the environmental impact assessment indicators when renovating a residential neighbourhood?**

References to Relevant Research

The concept of obtaining a system capable of helping us assess and minimise the environmental impact associated with buildings that form cities is not an isolated line of thought since other researchers have developed studies whose purpose is related to city sustainability and its environmental impact. For instance, the work entitled *Report on Sustainability Indicators (Informe sobre los indicadores de sostenibilidad)* [6] strives to conduct an analysis of the process developed by each municipality, and of the control indicators used when signing the Aalborg Charter. For this purpose, a survey was conducted on a selection of Spanish municipalities that had signed (as of June 2002) the Aalborg Charter, in order to make an approach towards the development of sustainability performed by these municipalities. The actual indicators used by 30 of these towns are collected, thereby accumulating a selection of 50 basic indicators provided in file cards in the document.

Bioclimatic Urbanism [7] describes the historical background and current state of the cities, the bioclimatic urban methodology, and planning according to bioclimatic urbanism principals. Finally, urban design criteria are presented. To this end, it is proposed that each place defines its own urban planning by structuring: road paths that meet sunlight and wind criteria (hierarchy and cross section); streets adapted to topography, seeking optimal sunlight orientation and wind orientation; urban garden structure appropriate to the needs of humidity and open-air evaporation (free surface and appropriate plant species); urban block morphology and subdivisions that create well-oriented façades and courtyards of an appropriate proportion to the climate; and diverse building typologies appropriate to the local sun and wind conditions.

Finally, "*The experience of rehabilitating homes in Spain using environmental parameters. The state of the art*" [8] is mentioned, which presents work expressly performed for a comprehensive urban renewal project in Palma de Mallorca. Under environmental parameters for this work, various actions to rehabilitate buildings are analysed (mainly Public Promotion Housing) in Spain in the 90s and 2000.

METHODOLOGY

The objective of the study is to define the assessment indicators for environmental impact in residential neighbourhoods, and to be able to determine the main factors when making their environmental rehabilitation.

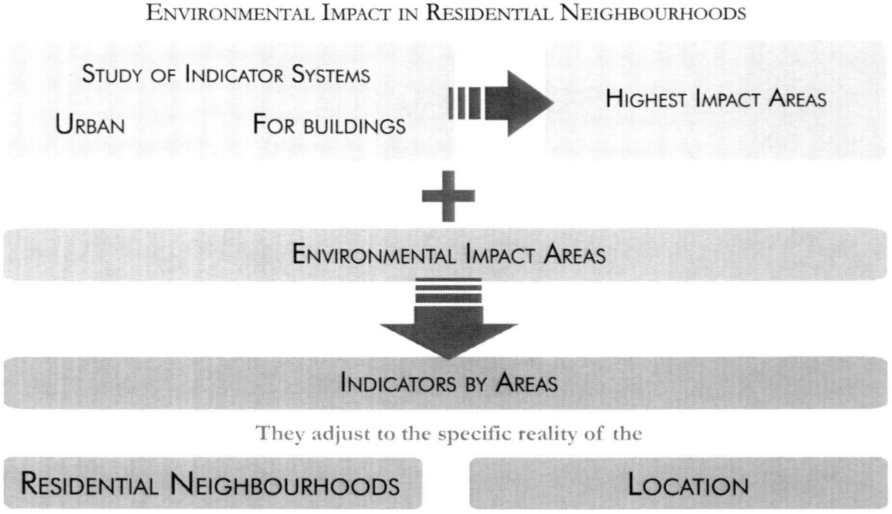

Figure 1: Process of the creation of Environmental Indicator System for Residential Neighbourhood Renovation.

An exhaustive study of the indicators is therefore used in current programmes and systems. Moreover, the indicator systems of the assessment method are analysed. For this reason, the first task is to establish a classification of these indicators.

Finally, the environmental impact indicators for residential neighbourhoods are determined, based on the areas of greatest impact on their assessment. Since the definition of the indicators of the new system is needed from a regulatory framework and the system needs to be adapted to the features of the location itself, these are also specified (Fig. **1**).

CLASSIFICATION OF INDICATORS FOR ENVIRONMENTAL REHABILITATION OF RESIDENTIAL NEIGHBOURHOODS

Indicators Regarding the Nature of their Data

On the one hand, the classification distinguishes between simple and complex indicators (also known as synthetic indicators or indexes) (Fig. **2**). The simple indicators refer to not very elaborate statistics directly obtained from reality, while the synthetic indicators are non-dimensional measures as a result of combining various simple indicators, by a weighting scheme which puts the components into a hierarchy. Greater information is obtained from these indicators. Nevertheless, its interpretation is more difficult on many occasions, and has certain restrictions [9].

On the other hand, we differentiate between quantitative, mixed, and qualitative indicators (Fig. **2**). This depends on the information they provide: values that express quantity or frequency of a fact or phenomenon (quantitative indicators); quantification of the quality of a phenomenon (mixed indicators); or they point towards a quality of the phenomenon by characterising it (qualitative indicators).

Indicators Regarding the Evaluation Goal

These can be distinguished between environmental sustainability indicators and environmental impact indicators (Fig. **2**).

An environmental indicator is an environmental factor that transmits information about the ecosystem of which it forms part. The environmental sustainability indicators allow many of the aspects that determine the sustainable level of an element to be synthesised into figures, thereby taking a more sustainable model as reference.

The environmental impact indicators are those environmental indicators that are employed to determine the environmental quality or its change when associated to a specific action [10]. In other words, it is an estimation of the magnitude of a given environmental impact as the difference in quality produced in the environmental factor affected.

Indicators Regarding the Subject of the Evaluation

In the case of the "Barriadas", we distinguish between urban indicators and indicators for buildings according to the nature of the subject, while bearing in mind that residential neighbourhoods comprise buildings that constitute urban fabrics (Fig. **2**).

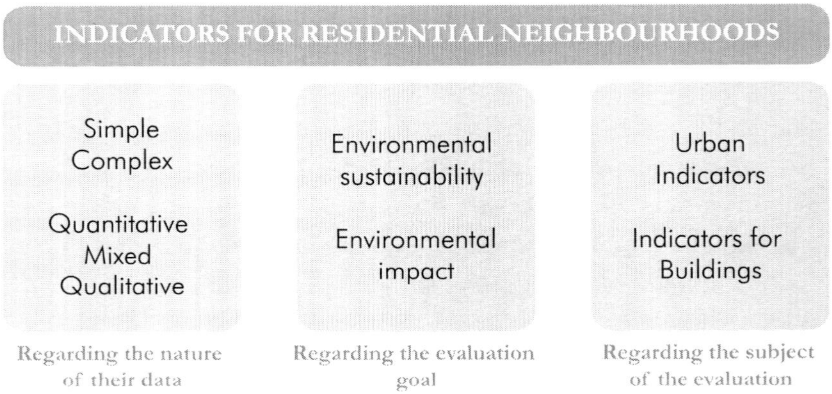

Figure 2: Classification of indicators for Residential Neighbourhoods.

The urban indicator is a variable or urban estimation that provides aggregated information: it is a complex or synthetic indicator that enables us to ascertain, in greater depth, the complex reality of the town/city.

The indicators for buildings permit the evaluation of the building, and determine the impact and aspects of the building and its plot. Thus, the client, user, and architect are allowed to make decisions and select alternatives that promote the necessary sustainability of buildings [11].

ENVIRONMENTAL ASSESSMENT INDICATOR FEATURES AND ENVIRONMENTAL IMPACT INDICATOR FEATURES OF RESIDENTIAL NEIGHBOURHOODS

To evaluate residential neighbourhoods, the indicators used must be able to reflect their dual reality: on the one hand, this reality is related to their nature as buildings, but on the other hand, these buildings are part of a larger element. Therefore, the indicators applied are simple and complex; quantitative and mixed, urban and for buildings.

These indicators, developed in the new system, include information on, but not limited to, the following:

- Typology of the building.
- Profile of use.
- Relevant technical and functional requirements.
- Required service life.

In addition, the indicators that allow us to assess the environmental behaviour of the neighbourhoods should apply measures related to aspects and environmental impacts of the building during its life cycle. This assessment should be based on life cycle analysis following the guidelines and requirements, as laid out in EN ISO 14044, and other quantifiable environmental information [12].

Case Study: Barriada Huerta del Carmen, Seville

In performing this research, a representative case study of this phenomenon has been chosen, "Barriada Huerta del Carmen", a development of 805 households built between 1958-1962, which is made up of 5-storey H-shaped building typologies and linear blocks, located in the Macarena district in Seville (Spain) (Fig. **3** and Fig. **4**).

The public space and green areas are mainly located in the residual spaces between the blocks. It is a generally modest housing construction.

"Barriada Huerta del Carmen" is a working-class neighbourhood inhabited by workers who arrived in the capital of Andalusia during the 60s and 70s and it presents a characteristic urban image of the "workers' quarters" of this period. It is perceived as the zone of greatest social vulnerability of the area due to its poor housing conditions and special de-structuring of public space [13].

This housing estate was built on the northern extension of Seville, comprising a group of 805 subsidised households, a school block, a parish centre (including the church of San Leandro) and a shopping arcade, and was developed by the Royal

Board of Cheap Houses (Real Patronato de las Casas Baratas) between the years 1958-62, and designed by the architect Fernando Barquín.

Figure 3: Barriada Huerta del Carmen. Building site [14].

The northern extension of land of Seville is nowadays structured by this kind of neighbourhood unit. This occurred due to the possibility of creating a higher division of the property in this part of the city, with the land organised in vegetable gardens. Moreover, the great accessibility to the extensive network of roads, as well as the immediate connection with the "Ronda de la Muralla" sector and the historical city centre itself, facilitated the replacement of the existing gardens with the new neighbourhoods. "Barriada Huerta del Carmen" was built on vegetable gardens located just inside the perimeter boundary wall built for the defence against floods during the Universal Exhibition of 1929.

The "Barriada" is the result of combining double centreline blocks to achieve urban order, whereby the design emerged as the integration of the housing architecture and the urban design; in other words, it is the result of the conjunction of the principles which conceive the city form as architecture. The internal organisation of the neighbourhood is completely independent of the main transit articulation of the city [15].

Figure 4: Barriada Huerta del Carmen at the present time.

INDICATOR SYSTEM ANALYSIS

We proceed to study in depth indicator systems currently used in the assessment of sustainability and environmental impact. Firstly, urban indicator systems (Table **1**) are analysed, indicating the areas of greatest impact. Then, we study systems of sustainability indicators for buildings (Table **2**), again analysing their areas of greatest impact.

The studied systems selection is based on the following factors:

- Influence
- Adaptation in the case of residential neighbourhoods
- Adaptability to the case study

These indicators are hereby related

Table 1: City level: urban sustainability indicators

City level: urban sustainability indicators
ISO 37120:2014 (en) Sustainable development of communities — Indicators for city services and quality of life.
White Book of Sustainability for Spanish Urban Planning.
Bases for an Urban Environment Indicator System in Andalusia. International experiences in the sustainable measurement in towns.
System of indicators and conditioning factors for large and medium-sized cities.
Municipal System of Sustainability Indicators.
Special Plan of indicators of environmental sustainability of the urban activity of Seville.
Indicators for urban sustainability in Vitoria-Gasteiz.

Table 2: Building level: building construction sustainability indicators

Building level: building construction sustainability indicators
Indicator system of sustainability in residential buildings for Andalusia.
Guíde to sustainable building for residential buildings in the Basque Country.
Environmental assessment methods: BREEAM, LEED, CASBEE, SbTOOL, Certification VERDE, Demarchè HQE, Green Star, DGNB, CFSH, Green Globes, ITACA Protocol, BEAM.

Evaluation Areas in the Urban Sustainability Indicator Systems Studied and in the Systems of Indicators for Buildings Studied

The main evaluation areas of the indicator systems under study are shown (Fig. **5**).

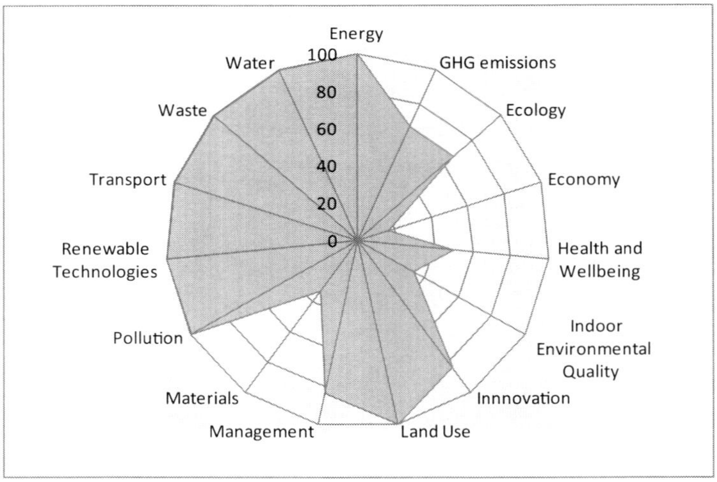

Figure 5: Evaluation criteria for urban indicator systems.

Among the systems analysed, 100% of the following areas are taken into account: energy, land use, pollution, renewable technologies, transport, waste, and water.

It can therefore be determined that the target of the indicator systems under study is to reduce energy consumption in buildings and in public spaces, to reduce the energy consumption derivative of urban mobility, and to facilitate waste management and the management of the water cycle.

Regarding the systems of indicators for buildings, the main areas of evaluation are listed below (Fig. **6**).

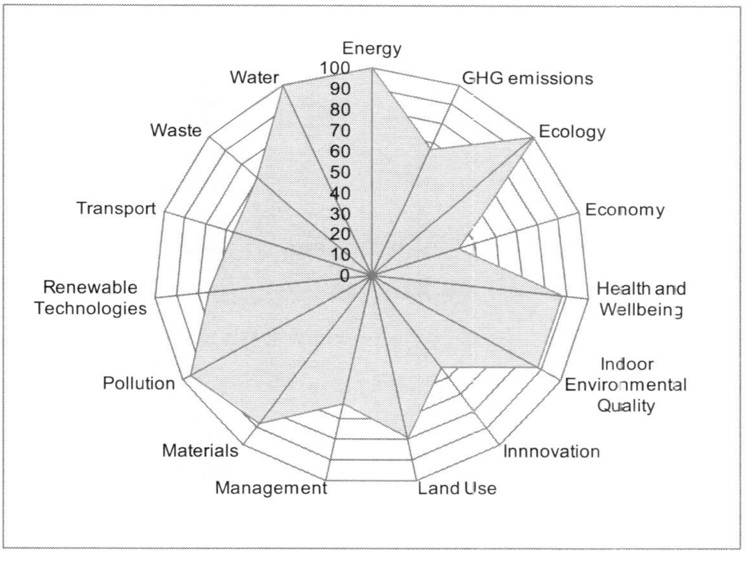

Figure 6: Evaluation criteria for sustainability indicator systems in residential building.

In this case, the areas with the greatest impact are energy, ecology and water, being followed by health and wellness, indoor air quality, materials and pollution.

Lessons Learned from the Study of Current Systems of Indicators as a Basis for Preparing Environmental Impact Indicator Systems in Residential Neighbourhood Renovation

A number of the areas of sustainability in building evaluation match those for urban sustainability. These are the cases in the areas of energy, pollution and water. However, it is important to emphasise the fact that the areas of importance when assessing the sustainability of the city (land use or transportation, for example), cease to be significant in the evaluation of buildings, mainly due to the change of scale, and other areas, such as ecology, health and wellness, indoor air quality, and materials rise in importance.

In addition, the energy consumption associated to buildings can be divided into three sections: energy use (active functions), the energy associated with the structure of the building, and the related mass flow (non-energetic material flow that enters and leaves the building: water and waste). Therefore, new developments and urban recycling processes should incorporate these criteria to encourage savings and efficiency. Saving involves giving up the use of energy resources that are not strictly necessary to cover certain basic functions, while maximising efficiency refers to the provision of a service (heating, cooling and lighting) with minimal consumption.

Regulatory Framework and Local Characteristics for the Development of Indicators

For the definition of the indicator system, its regulatory framework is studied at European, national and local levels, while considering the legislation relating to energy, renovations and the environment. The local characteristics related to the location are also taken into account.

European Regulatory Framework

- Directive 2012/27/EU of the European Parliament and of the Council of 25 October 2012 on energy efficiency, amending Directives 2009/125/EC and 2010/30/EU and repealing Directives 2004/8/EC and 2006/32/EC [16]: this aims to create a common framework to promote energy efficiency within the EU, by establishing specific actions to test some of the proposals included in the “Energy Efficiency Plan 2011” while achieving significant potential energy savings not yet performed.

- DIRECTIVE 2010/31/EU OF THE EUROPEAN PARLIAMENT AND OF THE COUNCIL of 19 May 2010 on the energy performance of buildings [17]: this aims to promote energy efficiency in buildings, and the integrated energy performance of buildings or building units; where the energy efficiency of building is “the calculated or measured amount of energy needed to meet the energy demand associated with a typical use of the building, which includes, among other things, energy used for heating, cooling, ventilation, hot water, and lighting” [17].

By this standard, the existing buildings, when undergoing major renovation works, should benefit from improved energy efficiency so that they can also fulfil the minimum requirements.

- Europe 2020 strategy; "A strategy for smart, sustainable and inclusive growth" [18]: it is proposed, as one of the five major objectives for the European Union, that the European Council adopted in March 2007, known as 20-20-20: "Reduce greenhouse gas emissions by at least 20% compared to 1990 levels, or by 30%, if the conditions are right; increase the share of renewable energy sources in our final energy consumption to 20%; and a 20% increase in energy efficiency" [18].

National Regulatory Framework

- Spanish Royal Decree 235/2013 "basic procedure for the energy certificate of buildings" [19]: It establishes the obligation to make an energy efficiency certificate available to purchasers or users of buildings, that should include objective information on the energy efficiency of the building and reference values such as energy performance minimum requirements to owners or tenants of the building or a unit of it to compare and assess its energy efficiency. The minimum energy efficiency of buildings or its units are not included in this Royal Decree, as they are set in the Technical Building Code (Código Técnico de la Edificación, CTE). Thus, by valuing and comparing the energy performance of buildings, it favours the promotion of energy-efficient buildings and investments in energy savings. Furthermore, this Royal Decree contributes towards reporting CO2 emissions from the use of energy from emission sources in the residential sector, which will facilitate the adoption of measures to reduce emissions and improve the energy rating of buildings.

- Spanish Royal Decree 233/2013, of 5 April, "regulating the State Plan to promote house renting, building rehabilitation and urban regeneration and renovation (2013-2016)" [20]: its objective is to regulate the state plan of development of rental housing, the rehabilitation of buildings and urban regeneration and renewal, 2013-2016. It includes the following programs:

  - Program of promotion of building rehabilitation.

- Program of promotion of urban regeneration and renewal.
- Program of support of the implementation of the evaluation report of buildings

- Spanish Law 21/2013, of December the 9th, on environmental assessment [21]: it aims to ensure the maximum environmental protection and give a new impulse to sustainable development. This regulation simplifies and streamlines the environmental assessment of plans, programmes and projects, since it is assumed that excessively lengthy procedures do not protect the environment and pose an obstacle to sustainable development. The undue delay of this procedure no longer protects the environment, but generates a negative image of the environmental regulations.

  It sets the primary obligation to submit any plan, program or project likely to have significant effects on the environment to an adequate environmental assessment before its adoption, approval or authorisation, or, if applicable, in the case of projects before filing an affidavit or prior notification referred to in Article 71 of Law No. 30/1992 of November 26, 1992 on General Government and the Common Administrative Procedure.

- Spanish Law 8/2013 of June 26 Rehabilitation, Regeneration and Urban Renewal [1]: this aims to enhance building refurbishment and urban regeneration and renewal, removing existing barriers and creating specific mechanisms to make it feasible and practicable; it provides an appropriate regulatory framework to enable the restructuring and revitalisation of the construction industry, finding new areas of activity, promoting quality, sustainability and competitiveness, both in buildings and on the land, bringing the Spanish regulatory framework closer to the European framework, particularly in relation to the objectives of efficiency, energy saving and the fight against energy poverty.

- Regulations for Thermal Installations in Buildings (in Spanish, RITE) [22]: it constitutes the basic regulatory framework within the requirements for energy efficiency and security by which thermal installations in buildings are regulated, in order to meet the demand for welfare and health of the people.

- Technical Building Code (in Spanish CTE) [23]: It is designed to improve the quality of the buildings, and promote innovation and sustainability. It is a policy instrument which sets the basic requirements of quality of buildings and facilities. Through this legislation, a certain satisfaction level is given to basic building requirements related to the safety and welfare of the people, which refers both to structural safety and fire protection, protection against noise, energy conservation or accessibility for people with disabilities.

- Spanish Law 38/1999 Building Act (in Spanish, LOE) [24]: its ultimate aim is to respond to the growing social demand for quality, by establishing the basic requirements that buildings must meet to ensure the safety of the people, the welfare of society and environmental protection. The main objective is to regulate the process of updating and completing the legal construction configuration of the agents involved, setting their obligations in order to establish responsibilities and covering the guarantees to the users, based on a definition of the basic requirements that buildings must meet.

Local Regulatory Framework

- Royal Decree 3/2015, March 3rd, which modifies Law 7/2007, July 9th, Law 7/2007 of June 9th, on the integrated management of environmental quality [25]. It includes the adjustment to the procedure Unified Environmental Authorisation (Autorización Ambiental Unificada, AAU), (not yet scheduled in the Spanish law), to the basic state legislation; changing the procedures for the environmental assessment of plans and programmes, which would be subject to Strategic Environmental Assessment, ordinary or simplified, according to a new relationship of plans and content types and the elimination of environmental qualification process to numerous activities of low environmental impact, replacing it with an affidavit for its promoters.

- Program to Promote Sustainable Construction in Andalusia [26]: it aims to facilitate energy savings and efficiency investments as well as those for renewable energy in buildings located in Andalusia, and the call has been scheduled for years 2014 and 2015.

Among the various lines of action, incentives are contemplated to promote actions that encourage energy savings, improved energy efficiency and the use of renewable energy in buildings located in Andalusia, through the completion of renewal works, refurbishment, suitability for use and efficient facilities. Moreover, the establishment of the Bureau of Construction and Sustainable Rehabilitation of Andalusia, regarded as a specialised area of participation in sustainable construction that defines and drives the Comprehensive Plan for Sustainable Construction and Rehabilitation of Andalusia in the period 2014-2020.

- Law 2/2007 for the Promotion of Renewable Energies and Energy Saving and Efficiency in Andalusia [27]: it facilitates the exploitation of the natural wealth of Andalusia in renewable energies through the organisation of more rational and appropriate use, as provided in Article 45 of the Spanish Constitution and the European, Spanish and Andalusian energy plan projects.

- Law 7/2007 of July 9th, on the Integrated Management of Environmental Quality [28]: it aims to establish an appropriate regulatory framework for the development of environmental policy of Andalusia, through instruments that ensure the incorporation of sustainability criteria.

Other regulations:

- Decree 6/2012 of January 17th, which approves the Regulation on protection against noise pollution in Andalusia.

- Environmental Security Ordinance Responsible for Noise and Vibration – Seville City Council.

- Decree 357/2010 of August 3rd, approving the Night Sky Quality Protection Regulations against light pollution.

Local Definitional Characteristics of the Indicators

Building Tradition

We focus on the structural characteristics of neighbourhoods, a phenomenon which refers to social housing estates that were built during the rule of Franco, between 1957 (the year of the creation of the Department of Housing) and 1975 [3], these constructions used simplicity and the repetition of elements "for greater economy but without stinting on the robustness and good structure" [29].

In the "Barriada Huerta del Carmen" case, the outer envelopes are divided into a ground floor composition of wall thicknesses of one foot and a half of regular bricks and one foot on the remaining floors. The floors are of reinforced ceramic and the rooftop composition is "a la Andaluza" (in the Andalusian style) with a slope formation made by lime concrete and cinder and flooring. Finally, the exterior trim is made of cement mortar, painted with lime [29].

The specifications state that "With great economy the intended beneficiaries will be left with the possibility of completing the work" [29].

Neighbourhoods of similar features, such as "El Tardón" located in Triana (western Seville), was built in 1952. Its supporting structure is formed by load-bearing walls of 1-foot-thick solid brick and reinforced ceramic floors. The rooftop composition is "a la Catalana" (in the Catalonian style) with partition walls and the roof eaves are ceramic tiles taken with mortars [3]. The residential neighbourhood was developed by the group of architects José Fonseca, Manuel Ruiz de la Prada, José Gómez Mesa, José María Rodríguez Cano, J. Piqueras, and M. Bastarreche.

In the case of other cities, we find San Martin de Porres in Córdoba (early 60s) [30], designed by architect Rafael de la Hoz. Structurally, the first promotion was built exclusively with bearing walls of one-foot thick. In later stages, concrete frames appeared in the inner bays, keeping the resistant brick walls only in façades. The floor slabs used were very light (14 cm deep) and the gabled orangery consisted of the same floor type and was finished with fibre cement sheets.

Local Characteristics

As in the case of the regulations, we establish a defining framework of the new system by studying the characteristics of the location of the residential neighbourhood. This will consist of the following elements: population characteristics of the location, local weather data and energy consumption in the city, public transport and waste collection.

- Population of Seville: The population of Seville in January 2014 was 696,676 inhabitants, of which the female population (52.5%) was slightly higher than that of men (47.5%) [31].

- Local climate: Seville has a tendency towards the Mediterranean climate, with some continental characteristics. With variable rainfall, hot dry summers and mild winters, the city has almost 3,000 hours of sunshine annually [32].

- Energy consumption: According to the National Institute of Statistics, the highest energy consumption in Andalusia in 2009 and 2011 was in the form of electricity, followed by gas [33].

- Public transportation: Public transportation is provided by Urban Transport Seville, SA Municipal (TUSSAM), established by the Hon. City of Seville in 1975. Its purpose is to manage, operate and administrate the urban public transport in the city of Seville, covering a population greater than 700,000 inhabitants, distributed over an area of 142 km^2, through a network of 43 daytime bus lines and 9 night lines [33]. The total length of the network exceeds 600 km, with a fleet of 377 buses and 4 trams (plus another 16 vehicles of contract lines).

In addition, Seville has more than 170 kilometres of bike lanes [34] and in 2013 had 260 bike stations of the public bike system in Seville, SEVICI [35].

- Waste collection: LIPASAM is the municipal public cleaning service of Seville city council, responsible for cleaning 1,077 km of roads, the collection of urban waste, and for its subsequent treatment to save resources and prevent environmental pollution. For this task LIPASAM has an average workforce of 1,607 men and women and 461 vehicles [36].

INDICATORS DEFINITION

It is established that urban-level sustainability indicators focus on the areas of energy, land use, pollution, renewable technologies, transport, waste, and water. On the other hand, the areas of greatest impact in the edification systems of indicators under study are energy, ecology and water, followed on a second level by health and welfare, interior air quality, materials, and pollution.

It is important to remember at this point that the objective is the evaluation of the environmental impact in urban entities, which can be seen as "the alteration of the quality of the environment produced by a human activity" [10].

Moreover, regarding environmental aspects, the imbalances in the cycles of water, energy, and materials, and the air, sound and light pollution [37] have been detected as a key feature of unsustainability in the urban systems. These cycles are also called urban ecological cycles or cycles of the urban ecosystem, and they differ from those that can appear in the natural environment. Due to their lineal (instead of cyclical) nature and to their imbalances, these cycles are one of the top concerns of the urban ecosystem [7]. Thus, the system of indicators proposed focuses on the following sections (Table **3**):

Table 3: Imbalances in urban fabric and indicators

IMBALANCE	SYSTEM OF INDICATORS - AREAS
Energy cycle imbalances	Energy Indicators (EN)
Water cycle imbalances	Water Indicators (WA)
Materials cycle imbalances	Materials and Waste Indicators (MAT and WAS)
Pollution (air, noise and light)	Atmosphere Indicators (ATM)

Now we are going to expand on the indicators belonging to each area of the new system. These indicators are the result of the study of the indicators of the selected areas of the systems previously analysed, with previous adaptation to the specific casuistic of the residential neighbourhoods and to the specific location of Seville (Spain).

For each system, the objective to be achieved with each indicator is established as well as the methodology employed to achieve these indicators/objectives.

Energy Area

The main problems associated with the energy cycle in the urban fabric are mainly due to two causes. The first is their inability to produce energy to supply their needs. Consequently, it produces a strong dependency on external sources. The second problem is based on the low efficiency of the energy consumption of the urban entities.

The Energy Area indicators of the system have the objective of reducing the characteristic energy consumption of the Residential Neighbourhood (RN), owing to its nature as part of the urban fabric, as well as its configuration by the residential building typologies.

The system strives to establish a state of the energy use, and a state of the art of efficiency, and aims to reduce the environmental impacts of the RN. To this end, two strategies have been established: the generation of renewable energies and the adoption of actions for energy saving and efficiency.

Energy Indicators

"Urban fabric consumption" indicator: The objective of these indicators is to determine the state of the energy consumption of the residential neighbourhood and to minimise it. To this end, they focus on the use of passive systems: orientation, the building design, the envelope configuration, solar protection, ventilation, and infiltrations, *etc*.

- EN01 Energy consumption by Typology
- EN02 Energy consumption by Residential Neighbourhood

"Original source of energy" indicator: The objective of these indicators is to promote the use of renewable energies.

- EN03 Renewable energy production by Typology
- EN04 Renewable energy production by Residential Neighbourhood

"Energy efficiency of buildings" indicator: The objective is to increase the energy efficiency of the current buildings to decrease the environmental impact.

- EN05 Overall energy performance of the building by Typology

Water Area

The main problems associated with the water cycle in the urban fabric are mainly due to either the overexploitation of water courses and underground water, or to the pollution resulting from waste dumped in the water.

Water Indicators

"Urban fabric consumption" indicator: The objective is to reduce the water consumption of the residential neighbourhood.

- WA01 Water consumption for domestic use by Typology
- WA02 Water consumption for domestic use by Residential Neighbourhood

"Water reutilisation" indicator: The objective of this indicator is to quantify and increase the use of treated wastewater, and to decrease pollution.

- WA03 Reuse of treated wastewater. Volume of reused water per capita

Materials and Waste Area

The main problems associated with the materials cycle in the urban fabric are mainly due to the pressure exerted on the environment for their production, as well as the production of urban solid waste and hazardous waste inherent to urban activity.

The Materials and Waste System of Indicators applies in the following cases: in the first case, to the materials resulting from the maintenance (and possible rehabilitation) of residential neighbourhoods. In the second case, the waste created from the use of the residential neighbourhood. The third and final case is the construction and demolition waste (CDW) as a result of the maintenance and renewal of the residential neighbourhood.

Material and Waste Indicators

"Current use of materials" indicator: The objective of these indicators is to increase the use of materials that produce less environmental impact.

- MAT01 Nature of the materials used by Typology
- MAT02 Nature of the materials used in the Residential Neighbourhood (urban design)

"Waste production" indicator: The objective of this indicator is to quantify and decrease the waste produced.

- WAS01 Volume of waste generated (per capita and composition) per year

"Location and allocation of items for waste collection" indicator: The objective of this indicator is to assess and improve the availability of elements for the user for selective waste collection as well as the accessibility of clean points.

- WAS02 Proximity to a waste collection point from the housing access point

- WAS03 Provision of waste collection systems

- WAS04 Proximity to recycling centres

"Classification and valuation of CDW" indicator: The objective of this indicator is to minimise CDW production and promote its reutilisation.

- WAS05 Valuation of construction and demolition waste

Atmosphere Area

The main atmospheric problems associated to urban fabrics that are discussed in this system of indicators which are related to the existence of substances foreign to nature, the result of the derived processes of urban development: contaminant emissions and increased temperature (air pollution); the increase in the natural brightness of the background sky as a result of light escaping from external lighting installations (light pollution); and the increase of noise levels that exceed nuisance rates (noise pollution).

The emissions addressed in this section refer to cases of emissions produced by the use of the urban fabric itself and the emissions due to the location of the urban fabric in an urban context.

Atmosphere Indicators

"Polluting gas emissions" indicator: The objective of these indicators is to evaluate and reduce the gas emissions that pollute the atmosphere.

- ATM01 Emission of greenhouse gases into the atmosphere by Typology

- ATM02 Emission of greenhouse gases into the atmosphere by Residential Neighbourhood

- ATM03 Emission of greenhouse gases into the atmosphere by Other Agents

"Light pollution" indicators: The objective of these indicators is to evaluate and reduce the effects of artificial light emissions to the atmosphere, in order to maintain the natural conditions for the benefit of the ecosystem in general, by promoting energy efficiency of outdoor lighting and preventing the intrusion of unnecessary artificial light into houses.

- ATM04 Unnecessary light emission by Typology
- ATM05 Unnecessary light emission by Residential Neighbourhood
- ATM06 Obstruction of solar radiation by trees on public road network
- ATM07 Natural lighting in the interior spaces of the building

"Noise pollution" indicator: The objective of this indicator is to evaluate and reduce the effects of noise emissions in order to improve the quality of life for residents of Residential Neighbourhoods.

- ATM08 Measurement of acoustic comfort

System of Indicators for the Environmental Impact of the Rehabilitation of Residential Neighbourhoods

Finally, a summary of the proposed system of indicators is given in Table **4**.

Table 4: System of indicators for the environmental impact of the rehabilitation of Residential Neighbourhoods

INDICATOR	SYSTEM OF INDICATORS
Energy Indicators (EN)	EN01 Energy consumption by Typology EN02 Energy consumption by Residential Neighbourhood EN03 Renewable energy production by Typology EN04 Renewable energy production by Residential Neighbourhood EN05 Overall energy performance of the building by Typology
Water Indicators (WA)	WA01 Water consumption for domestic use by Typology WA02 Water consumption for domestic use by Residential Neighbourhood WA03 Reuse of treated wastewater. Volume of reused water per capita
Materials and Waste Indicators (MAT and WAS)	MAT01 Nature of the materials used by Typology MAT02 Nature of the materials used in the Residential Neighbourhood (urban design) WAS01 Volume of waste generated (per capita and composition) per year WAS02 Proximity to a waste collection point from the housing access point WAS03 Provision of waste collection systems WAS04 Proximity to clean points

	WAS05 Valuation of construction and demolition waste
Atmosphere Indicators (ATM)	ATM01 Emission of greenhouse gases to the atmosphere by Typology ATM02 Emission of greenhouse gases to the atmosphere by Residential Neighbourhood ATM03 Emission of greenhouse gases to the atmosphere by Other Agents ATM04 Unnecessary light emission by Typology ATM05 Unnecessary light emission by Residential Neighbourhood ATM06 Obstruction of solar radiation by trees on public road network ATM07 Natural lighting in the interior spaces of the building ATM08 Measurement of acoustic comfort

THE IMPORTANCE OF THE SYSTEM OF ENVIRONMENTAL IMPACT INDICATORS PROPOSED FOR RESIDENTIAL NEIGHBOURHOODS RENOVATIONS. CONCLUSION AND DISCUSSION

Throughout this chapter, we developed a methodology that has allowed us to identify sustainability indicators to be considered in the renovation of Residential Neighbourhoods. To this end, firstly the importance has been established of forming a system of indicators for assessing and minimising the environmental impact on the residential neighbourhood renovations, due to the absence of a specific system for that purpose and the impact that these urban fabrics have on Spanish cities. Below, the types of indicators used in developing the new system are described, concluding that due to the particular characteristics of the urban fabrics which constitute residential neighbourhoods, these indicators are both simple and complex, quantitative and mixed, urban and for buildings.

We have chosen the case study, the Barriada Huerta del Carmen, which serves as a model where the proposed indicator system can be tested. We have established a policy and local feature framework that will allow the definition of the indicators.

Through this analysis and by comparing the most important areas in existing systems of urban indicators and building results studied in the most important cycles of the imbalances in the urban ecosystem, it has been concluded that there are four areas of the new system of indicators: energy, water, materials and waste, and atmosphere. Finally, the indicators for each area are established with the objectives pursued by each one.

The importance of the system developed here derives from a misinterpretation of the “Barriadas” as sets of buildings (architectural objects) that, by repeating one or more residential building typologies of collective housing, conforms residential complexes that can be observed as episodes within the city rather than independent elements. In other words, these phenomena are evaluated either for their urban character or for their construction character, but rarely as the interaction of these two realities.

The research here presented is part of a larger process, the doctoral thesis entitled "METHOD OF ASSESSMENT AND MINIMISING THE ENVIRONMENTAL IMPACT OF COLLECTIVE HOUSING: Application to the Barriada Huerta del Carmen, Sevilla", currently under development by the authors. The ultimate goal of this research is the proposal of the environmental rehabilitation of the case study and the proposal of eco-efficient solutions to reduce the environmental impact of this common urban fabric in Spanish cities.

ACKNOWLEDGEMENTS

The research team thanks the Royal Housing Foundation of Seville (Real Patronato de la Vivienda de Sevilla) for the information provided about the Huerta del Carmen neighbourhood as well as Planning Department of the Council of Seville for the information about the city of Seville.

CONFLICT OF INTEREST

The authors confirm that this chapter contents have no conflict of interest.

REFERENCES

[1] J. del Estado, "*Ley 8/2013, de 26 de junio, de rehabilitación, regeneración y renovación urbanas*". Spain: Agencia Estatal Boletín Oficial del Estado, June 2013. [Online] Available; http://www.boe.es/diario_boe/txt.php?id=BOE-A-2013-6938 [Accessed Mar. 16, 2015]

[2] IDAE, "*Análisis del consumo energético del sector residencial en España. Informe final*". Madrid: IDAE Instituto para la Diversificación y Ahorro de la Energía, July 2011. [Online] Available; http://www.idae.es/index.php/mod.documentos/mem.descarga?file=/documentos_Informe_SPAHOUSEC_ACC_f68291a3.pdf [Accessed Mar. 16, 2015]

[3] E. Valero Ramos, "*5 Barriadas residenciales en Sevilla. Estudio pormenorizado*". Sevilla: Dirección General de Arquitectura y Vivienda. Conserjería de Obras de la Junta de Andalucia, July 2007 [Online] Available; https://ws147.juntadeandalucia.es/obraspublicasyvivienda/davwebviv/Proyectos%20de%20Investigacion/34%20RECICLAJE%20DE%20BARRIADAS,%20UNA%20ALTERNATIVA%20SOSTENIBLE/03_analisis%20barriadas/Estudio%20Barriadas%20Sevilla_Isabel%20Diaz/Estudio%20Barriadas%20Sevilla%20140909.pdf [Accessed Mar. 16, 2015]

[4] J. Rubio del Val, "*La hora de la rehabilitación urbana sostenible en España*", Cambio Global España 2020/50. Sector Edificación, pp.182–193, 2010.

[5] M. López de Asiaín, "*Indicadores de sostenibilidad, salvación o nueva pesadilla*", Sevilla: La Ciudad Viva, March 2010 [Online] Available; http://www.laciudadviva.org/blogs/?p=4845 [Accessed Mar. 16, 2015]

[6] Grupo de trabajo sobre los Indicadores de Sostenibilidad para la aplicación del Programa Hábitat "*Informe sobre los indicadores de sostenibilidad*", Madrid: Dirección General de la Vivienda, la Arquitectura y el Urbanismo, Ministerio de Fomento, 2004.

[7] E. Higueras, "*Urbanismo Bioclimático*", Barcelona: Editorial Gustavo Gili, 2006.

[8] X. Prat amd G. Wadel, "*La experiencia de rehabilitación de viviendas en España bajo parámetros ambientales. Un estado del arte*". Madrid: SB10mad, 2010. [Online] Available; http://www.sb10mad.com/ponencias/archivos/b/B014.pdf [Accessed Mar. 17, 2015]

[9] Junta de Castilla y León, "*Introducción*", 2014 [Online] Available; http://www.jcyl.es/web/jcyl/Medio Ambiente/es/Plantilla100/1218521583895/_/_/_ [Accessed Mar. 16, 2015]

[10] A. Garmendia, A. Salvador, C. Crespo and L. Garmendia, "*Evaluación de Impacto Ambiental*", Madrid: Pearson-Prentice Hall, 2006.

[11] AENOR, "*UNE-EN 15643-1 Sostenibilidad en la construcción. Evaluación de la sostenibilidad de los edificios. Parte 1: Marco general*", Madrid: Norma Española, 2012.

[12] AENOR, "*UNE-EN 15643-2 Sostenibilidad en la construcción. Evaluación de la sostenibilidad en los edificios. Parte 2: Marco para la evaluación del comportamiento ambiental*", Madrid: Norma Española, 2012.

[13] A. Hernández Aja, M. Vázquez Espí, C. García Madruga, A. M. Parellada, E. M. Garcia, J. A. Gómez, J. C. Gutiérrez. "*Análisis Urbanístico de Barrios Vulnerables*". Madrid: Departamento de Urbanística y Ordenación del Territorio/IJH, February 2013. [Online] Available; http://habitat.aq.upm.es/bbvv/ [Accessed Mar. 16, 2015]

[14] Equipo Barquín y Barón. "*Archive*". Seville: Real Fundación Patronato de la Vivienda, 1956.

[15] A. Barrionuevo Ferrer, "*Fernando Barquín y la construcción de la ciudad. Sevilla 1945 – 1965*", Fernando Barquín y Barón – Joaquín Barquín y Barón. Imágenes de su archivo en los fondos FIDAS, Sevilla: FIDAS, 2007.

[16] The European Parliament and the Council of the European Union, "*Directive 2012/27/EU of the European Parliament and of the Council of 25 October 2012 on energy efficiency, amending Directives 2009/125/EC and 2010/30/EU and repealing Directives 2004/8/EC and 2006/32/EC Text with EEA relevance*". Official Journal of the European Union, 2012 [Online] Available; http://eur-lex.europa.eu/legal-content/EN/TXT/?qid=1399375464230&uri=CELEX:32012L0027 [Accessed Mar. 16, 2015]

[17] The European Parliament and the Council of the European Union, "*Directive 2010/31/eu of the european parliament and of the council of 19 May 2010 on the energy performance of buildings*". Official Journal of the European Union, 2010 [Online] Available; http://europa.eu/legislation_summaries/internal_market/single_market_for_goods/construction/en0021_en.htm [Accessed Mar. 17, 2015]

[18] Europe Direct, "*Energy performance of buildings*". Summaries of EU legislation, 2010. [Online] Available; http://europa.eu/legislation_summaries/internal_market/single_market_for_goods/construction/en0021_en.htm [Accessed Mar. 17, 2015]

[19] Ministerio de la Presidencia, "*Real Decreto 235/2013, de 5 de abril, por el que se aprueba el procedimiento básico para la certificación de la eficiencia energética de los edificios*". Madrid: Boletín Oficial del Estado, 2013

[20] Ministerio de Fomento, "*Real Decreto 233/2013, de 5 de abril, por el que se regula el Plan Estatal de fomento del alquiler de viviendas, la rehabilitación edificatoria, y la regeneración y renovación urbanas, 2013-2016*". Madrid: Boletín Oficial del Estado, 2013

[21] Jefatura del Estado. "*Ley 21/2013, de 9 de diciembre, de evaluación ambiental*". Madrid: Boletín Oficial del Estado, December 2013.

[22] Ministerio de la Presidencia, "*Real Decreto 1027/2007, de 20 de julio, por el que se aprueba el Reglamento de Instalaciones Térmicas en los Edificios*". Madrid: Boletín Oficial del Estado, July 2007.

[23] Ministerio de Vivienda, "*Real Decreto 314/2006, de 17 de marzo, por el que se aprueba el Código Técnico de la Edificación*". Madrid: Boletín Oficial del Estado, March 2006.

[24] Jefatura del Estado, "*La Ley 38/1999 de 5 de noviembre, de Ordenación de la Edificación*". Madrid: Boletín Oficial del Estado, November 1999.

[25] Consejería de la Presidencia, "*Decreto-ley 3/2015, de 3 de marzo, por el que se modifican las Leyes 7/2007, de 9 de julio, de gestión integrada de la calidad ambiental de Andalucía, 9/2010, de 30 de julio, de aguas de Andalucía, 8/1997, de 23 de diciembre, por la que se aprueban medidas en materia tributaria, presupuestaria, de empresas de la Junta de Andalucía y otras entidades, de recaudación, de contratación, de función pública y de fianzas de arrendamientos y suministros y se adoptan medidas excepcionales en materia de sanidad animal*". Sevilla: Boletín Oficial de la Junta de Andalucía, March 2015.

[26] Consejería de Economía, Innovación, Ciencia y Empleo, "*Decreto-ley 1/2014, de 18 de marzo, por el que se regula el Programa de Impulso a la Construcción Sostenible en Andalucía y se efectúa la convocatoria de incentivos para 2014 y 2015*". Sevilla: Boletín Oficial de la Junta de Andalucía, March 2014.

[27] Presidencia de la Junta de Andalucía, "*Ley 2/2007, de 27 de marzo, de fomento de las energías renovables y del ahorro y eficiencia energética de Andalucía*", Madrid: Agencia Estatal Boletín Oficial del Estado, March 2007.

[28] Presidencia de la Junta de Andalucía, "*Ley 7/2007, de 9 de julio, de Gestión Integrada de la Calidad Ambiental*". Madrid: Agencia Estatal Boletín Oficial del Estado, July 2007.

[29] F. Barquín, "*Proyecto de 805 viviendas sociales y locales comerciales en la Barriada del Carmen (Macarena). Memoria*". Sevilla: Estudio Barquín y Barón, 1959.

[30] J. Benítez-Castro, "*Experiencia en barrios: San Martín de Porres, Córdoba*", La Ciudad Viva, vol. 03, pp. 46-53, 2009.

[31] INE, "*Instituto Nacional de Estadística*". Madrid: Instituto Nacional de Estadística, 2015. [Online] Available; http://www.ine.es/en/welcome.shtml [Accessed Mar. 16, 2015]

[32] Ayuntamiento de Sevilla, "*Climate*". Sevilla: Ayuntamiento de Sevilla, 2015. [Online]. Available; http://www.sevilla.org/city/climate [Accessed Mar. 16, 2015]

[33] TUSSAM, "*Medio ambiente*". Sevilla: Ayuntamiento de Sevilla, 2014. [Online]. Available; http://www.tussam.es/index.php?id=20 [Accessed Mar. 16, 2015]

[34] I. Valdés Aragonés, "*Sevilla, paraíso de las dos ruedas*", El País, May 11, 2014 [Online] Available; http://elpais.com/elpais/2014/05/07/eps/1399465309_221018.html [Accessed Mar. 16, 2015]

[35] Servicios de Proyectos y Obras – Sección Técnica de vías ciclistas, "*Estaciones Del Servicio Público De Bicicletas Sevici – Año 2013*". Sevilla: Gerencia de Urbanismo, 2013. [Online] Available; http://www.sevilla.org/sevillaenbici/Contenidos/1-enbici/SEVICI%E2%80%93Estaciones 2013.pdf [Accessed Mar. 16, 2015]

[36] LIPASAM, "*Informes y estadísticas*". Sevilla: Ayuntamiento de Sevilla, 2015 [Online]. Available; http://www.lipasam.es/index.php?id=14 [Accessed Mar. 16, 2015]

[37] J.M. Castro-Bonaño and E. Salvo-Tierra, Eds., "*Bases para un Sistema de Indicadores de Medio Ambiente Urbano en Andalucía. Experiencias internacionales en la medición de la sostenibilidad en las ciudades*", Sevilla: Consejería de Medio Ambiente. Junta de Andalucía, 2001.

Chapter 3: Approach to the Definition of Nearly Zero Energy Buildings, in the Social Housing in Europe

CHAPTER 3

Approach to the Definition of Nearly Zero Energy Buildings, in the Social Housing in Europe

Begoña Serrano Lanzarote*, Alejandra García-Prieto Ruiz, Leticia Ortega Madrigal, Laura Soto Francés, Violeta de la Fuente Pérez

IVE Valencia Institute of Building, Retrofit Department, C/ Tres Forques, 98, 46018, Valencia, Spain

Abstract: This chapter presents the results of the Power House NZEB Challenge project [1], subsidize with funds from the Intelligent Energy Europe program, leaded by the public entity Housing Europe. In this context, different public developers of social housing are working on the definition of the concept and the design of Nearly zero energy buildings (NZEB), according to the exigencies present on article 9 of the 2010/31/EU Directive of Energy Efficiency of Buildings. The primary objective of the project is to transfer and exchange experiences between different countries about the design of this type of buildings, and inform to political lawmakers about the results. The project is studying thirty different buildings. Following the optimal cost methodology, as it is presented on the European Directive 2010/31/EU, inside its Delegate Norm 244/2012/EU tries to guarantee a rational use of the financial resources and not letting the economical investment to be disproportionate. This article presents one of the study cases from Spain, showing the results of applying the methodology of the optimal cost, which allows identify those technical solutions with the higher economical profitability considering the global cost of the next 30 years. Furthermore the problems that a study of these characteristics, apparently of easy implementation, usually shows have been identified. The knowledge generated will help to define the NZEB concept and will be useful as a guide for the Member States on the elaboration of the regulatory and financial framework; to assure that energy transition is inclusive and socially, economically and environmentally sustainable.

Keywords: Case study, cost-effectiveness, economical sustainability, energy efficiency, energy qualification, energy retrofit, energy transition, EPBD, environmental sustainability, European policies, global costs, Housing Europe, IEE programme, intelligent energy, NZEB building, NZEB concept, optimal cost, power house, social housing, standard.

***Corresponding author Begoña S. Lanzarote:** IVE Valencia Institute of Building, Retrofit Department, C/ Tres Forques, 98, 46018, Valencia (Spain); Tel: +34 961207531; Fax: +34 961207542; E-mail: bserranol@five.es.

Pilar Mercader-Moyano (Ed.)

INTRODUCTION

nZEB Normative Framework

The 2010/31/EU Directive of Energy Efficiency of Buildings [2], in its article no. 9, about Nearly zero energy buildings, indicates that the Member States will ensure that no later than December 31st of 2020, all the new buildings constructed must be, at least, nearly zero energy buildings. Spain, so far, does not have defined the NZEB concept that will specify the exigencies and requirements to be comply by new buildings after 2020.

In its article no. 2 defines, among others, the concept of "building of nearly zero energy" as a building with a really high level of energy efficiency. The quantity near zero or really low needed, will be supplied, mostly, by energy coming from renewable energies, including energies coming from renewable sources generated "*in situ*" or on the surroundings. It also defines the concept of "optimal level of profitability" as the level of energy efficiency with the lowest cost during its estimated life cycle, considering the investment costs related to energy, maintenance and working and the removal costs. The optimal level of profitability will be situated on the performance level ranges where balance cost-benefit calculated during the estimated life cycle is positive.

Description and Objectives of the Project Nearly Zero Challenge

The European Commission, in its intention of promoting energy savings on the civil sector, supports the project Nearly Zero Challenge [1], englobed inside the platform Intelligent Energy and coordinated by CECODHAS Housing Europe [4], with an important network of 42 national and regional associations that group together around 41.400 developers, in 22 countries, and manage more than 25 million houses in Europe. The rest of partners are formed by the housing associations of each country. In Spain, the entity is AVS [5] which works together with the Valencian Institute of Buildings (IVE).

From the work of the Member States on the definition and regulation of the "nearly zero energy buildings", CECODHAS faces the project as a network for the exchange and transmission of knowledge between the agents of social housing all around Europe. The project committee works on four areas and the work here presented belongs to the nearly zero energy buildings in the Mediterranean climate area, whose entity in charge of the coordination of the group team about the Mediterranean climate are AVS [5] and FEDERCASA [6].

The project aims to help its members on its way towards sustainability of buildings offering an international platform for the exchange of good practices, through thematic seminars, visits of studying, on-line data bases,... and a wide range of tools/information to be used by the National Associations to help their governments to define their housing strategies.

The specific objectives for the Mediterranean climate area are the following ones:

- It is intended to help to **define the concept of NZEB on Mediterranean climates**, attending to technical solutions, characteristics related to culture, traditions, climate... According to the document *Climate Change: Implications for Buildings* [7], the technical codes of construction and normative are, generally, a good solution to achieve CO2 emissions reduction, in an efficient, realistic and sustainable manner.

- **Case studies will be analysed**, to collect the direct opinion of the developers of these experiences, to identify the barriers that difficult its implementation and extract the lessons learnt from it that may help to define this concept and normative, based on verified and real information. Related to this, one of the works being developed inside this group, in collaboration with the Valencian Institute of Buildings (IVE), is **analyse and monitor the energy consumptions** of five buildings, whose design may be near the definition of "nearly zero energy building" considered in the European Directive [2] and yet pending to be defined on our country. The selection of these buildings is motivated because they were rated with the energy letter A and considering that this condition can be a starting point for the definition searched.

- To go in depth on the definition, it is intended to apply to the selected buildings the optimal cost methodology [8], following the comparative methodology guideline regulated on the Annex III of the Delegated Regulation [3]. It is been elaborated more specifically some reports about optimal cost, from the global costs on a 30 years term, considering maintenance, investment, work conditions, replacements, energy costs, ...

OBJECTIVES AND METHODOLOGY

This document describes the process followed to apply the methodology of the optimal cost to the Salburua building case of study in Vitoria-Gasteiz, which is part of a social housing project that is analysed by the Power House NZEB Challenge project [1]. The final results generate an interesting knowledge to help to define the concept of NZEB buildings in Mediterranean countries and in the social housing sector. Due to the inexistence of a Technical Code of Building with a NZEB bases in Spain, the criteria followed to select the analysed buildings have been to consider those buildings with an A energy certification. On 2012, starting year of the project, only 3 public development housing buildings were identified as a letter A that had already been completed. It is important to do these studies with an economic efficiency point of view and not just about energy efficiency. They should consider simultaneously environmental aspects (reduction of energy consumption and CO2 emissions), social and economic (investment, energy price, maintenance costs…) to assure an inclusive and sustainable energy transition to NZEB buildings [9].

The methodology tries to analyse a building with different energy improvement options, to be able to compare and establish the most optimal solution from an economical profitability point of view, considering the global cost for a 30 year period. Therefore, it is a methodology that simultaneously considers environmental as well as economic and social aspects.

A package of 9 technical improvements measures have been suggested, grouped by each of the three normative categories considered. The first have been established following the requests needed to comply the document of energy savings of the Technical Code of Construction of 2006 (CTE06) [10]. The second with the update that was approved in 2013 (CTE13) [11] and, the last one, considering that a "nearly zero energy building" (NZEB) is equivalent to an energy letter A. Each one of the global costs on a 30 years term, of each one of the proposed solution packages, were evaluated according to the phases described on the Delegate Regulation [3] and following the methodology about optimal cost developed by the BPIE [8]. The global costs include the construction costs, maintenance costs, the costs of renovating the construction elements that need to be replaced during the observing period, as well as the energy costs.

CHARACTERISTICS OF THE STUDIED BUILDING

The Salburua study case building is newly constructed and its final use will be residential with 242 social houses. It has been developed by VISESA, public developer institution of the Basque government.

Characteristics of the Site

The building is situated on the left side of the Salburua Boulevard, at the east area of the city of Vitoria-Gasteiz in Spain (Fig. **1**). It is an expansion area growing on the east side of Vitoria-Gasteiz, specifically on the Sector 11, plot M3, of Salburua.

a) b)

Figure 1: Situation: a) City; b) Area in the city. (Source: Google maps).

The city of Vitoria-Gasteiz is situated inside the D1 climate area, with an altitude from sea level of 512 m. Its climate characterizes by maximum temperatures of 27°C during the summer and 1°C during winter.

Architectonical and Urbanism Characteristics

The site, according to the criteria established on the urban planning (Fig. **2**), is occupied by a tower of 21 floors and a continue block with U shape and variable height, ranging from four to seven floors. At ground level are situated the entrances to the building and nine commercial spaces. The two underground floors occupy the entire plot and are used as parking spaces, technical rooms and storages of the tower; and in the between roofs floor, the storages of the rest of houses. The tower has a southwest orientation, and the apartments area with a U shape, a north, east and west orientation takin the urban grid as reference.

The next pictures show a transversal picture (Fig. **3**) and the standar floor plant of the building (Fig. **4**). The houses are designed with two façade, situating the bedrooms next to the interior garden patio façade while the living rooms and the kitchen are next to the street façade. Doing so, a crossing air ventilation is guatanteed. The west-east façade has been constructed as a wall with narrow windows with horizontal proportions included on a continuous metal sheet wall. The north façade copies the design but with a more hermetic appearance. The south façade works as a collector of solar radiation, its gaps are wider, from ground to ceiling, protected with shutters to mitigate the heat lose during the night. The south façade has been provided with terraces that work as blinds for solar protection during the summer.

Figure 2: Urban planning. (Source: VISESA).

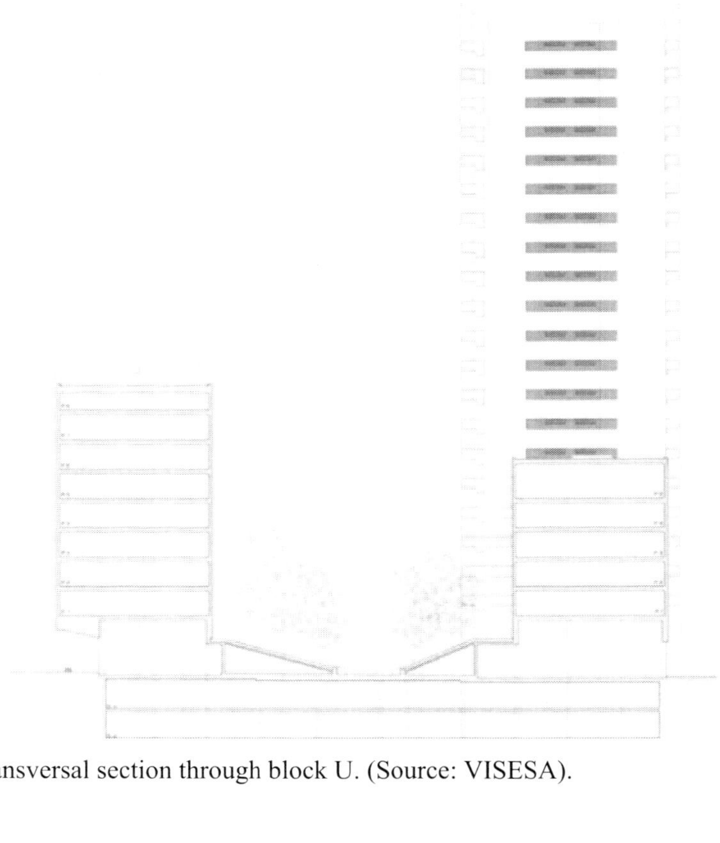

Figure 3: Transversal section through block U. (Source: VISESA).

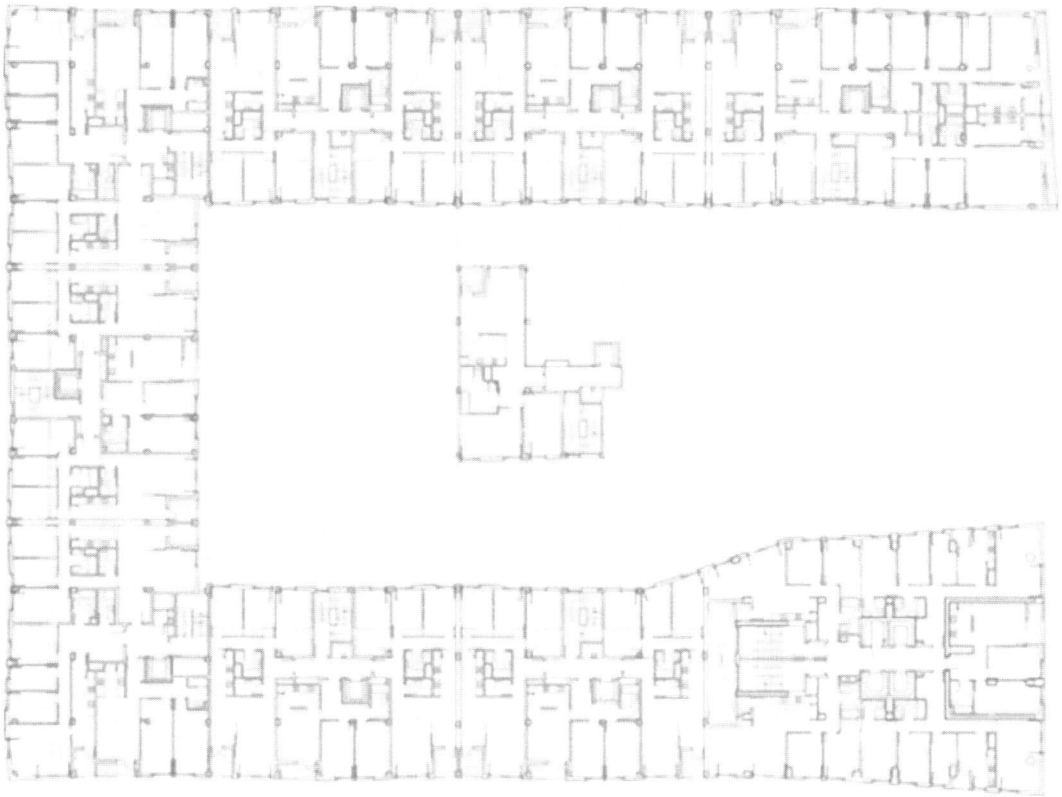

Figure 4: Type floor. (Source: VISESA).

Resume of the characteristics of the example building (Table **1**).

Table 1: General characteristics of Salburua building

Type of project:	New construction
Type of building:	Apartment block with tower
No. of houses:	242
No. of floors:	20 floors tower and block between 4 and 4 floors
Total area:	20.881,70 m^2
Total constructed area:	18.117,78 m^2
Ownership:	Private property
Energy qualification:	A
Actual occupancy rate:	73%
Architect:	ACXT Architects
Construction end date:	2011

Construction Characteristics of the Thermal Envelope

Façades

The façades are constructed with an interior wall of self-support gypsum and carton, with glass fibre insulation of 50 mm, a wall of ½ foot of perforated brick, a layer of water-repellent mortar, a continuous insulation of rockwool 80mm (40+40), and a exterior layer of lacquer waving steel sheets fixed by Z beams of galvanize steel.

Roof

The roof is formed by a plane walkable inverted surface in the entrances 1 and 2, whose heavy protection will be created with boulder gravel. Over the rest of the entrances, a plane no walkable inverted roof will be constructed with the same materials as the façades. In the underground floor are situated most of the storage spaces, which helps to create a chamber between the houses and the exterior.

On the walkable roofs it is used expanded polystyrene of 100 mm as insulation, and on the no-walkable roofs a 40 mm of rockwool is added.

Windows

The windows are made from natural anodize aluminium with heat channel breakage. The glazing is formed by two layers of hermetic glass with low emission. In laboratory tests, done by the responsible entities of the Basque government, the total thermal transmission is: $U=2,55 \pm 0,13$ W/m^2K.

Air Conditioning Characteristics of the Building

The thermal installation is centralized and it has cogeneration equipment, fed by gas, which produces electricity that is afterwards sold to an electricity supplier. The burning process of the gas generates heat energy that is used to warm water that is accumulated into a deposit. The energy cogeneration satisfies a 100% of the heat water demand and the 11% of heating demand, which is completed with two low energy burning boilers. The cogeneration equipment can produce 70kW with energy regulation depending on the demand and the two boilers produce 895kW of power, with an accumulation capacity of 20000 litres. It does not have air conditioning system due to the climate conditions of Vitoria.

The installation has four close circuits: one for the heating installation for the entire building and three for the water heating system (one for the tower, another for the 1-2-3-4 entries, and the third for the 5-6-7-8-9 entries).

DESCRIPTION OF THE PROPOSED SOLUTIONS

In this work have been defined a total of 9 packages of solutions about the analysed building and have been quantified the energy consumption and demand. The final objective is to create a three categories of improvements, each one of them related to different normative periods linked to the Technical Code of Construction covering the years 2006 (CTE06) [10], 2013 (CTE13) [11] and which may be the normative in 2020 (NZEB), adding the criteria of nearly zero energy buildings. For this study has been considered that the model building would be one that gets an energy letter A.

About the Thermal Envelope

The criteria followed to establish the solutions for the thermal envelope, on the CTE 2006 [10] and CTE 2013 [11] scenarios, has been the compliment of regulations in each of the cases, regarding thermal transmission values as well as maximum values admitted. On the NZEV scenario, the criteria followed have been to achieve an energy A class regarding demand.

The technical versions of the different solutions are specified below:

Thermal transmission, U (W/m^{2}K), of the different opaque construction elements forming the thermal envelope (façades, roofs, dividing wall, floors, and windows). Different values have been assigned for the complying of the three scenarios CTE 2006 [10], CTE 2013 [11] y NZEB, as it is shown in Table **2**:

Table 2: Thermal transmission of the construction elements for each technical version

Combinations		U (W/m^{2}K) **façade**	U (W/m^{2}K) **roof**	U (W/m^{2}K) **dividing wall**	U (W/m^{2}K) **floor**
C1-C3	CTE06	0.66	0.21/0.41	1.00	0.41
C5-C7	CTE13	0.27	0.21/0.41	0.95	0.41
C9-C11	NZEB	0.20	0.21	0.95	0.41

Thermal transmission, U (W/m^{2}K), of the empty spaces of the thermal envelope. Different values have been assigned for the complying of the three scenarios CTE 2006 [9], CTE 2013 [10] y NZEB (Table **3**).

Windows permeability (m^3/hm^2) of the thermal envelope. Different values have been assigned for the complying of the three scenarios CTE 2006 [10], CTE 2013 [11] y NZEB (Table **4**).

Table 3: Thermal transmission of empty spaces for each technical version

Combinations assigned		**U glass (W/m²K)**	**U frame (W/m²K)**	**U gap (W/m²K)**	**Solar index (%)**
C1-C2-C3	CTE06	Low emission 4-6-4 (0.1-0.2) 2.70	Wood medium-low density 4.00	3.09	0.52
C5-C6-C7	CTE13	Low emission 4-12-331(0.10.2) 2.00	Metallic with heat channel breakage 4-12mm 4.00	2.60	0.52
C9-C10-C11	NZEB	Double low emission <0.03 4-15-6 1.40	PVC three chambers 1.80	1.50	0.54

Table 4: Windows permeability

Combinations assigned		**Windows permeability**
C1-C2-C3	CTE06	27 (swing, good fit with draught excluder)
C5-C6-C7	CTE13	27 (swing, good fit with draught excluder)
C9-C10-C11	NZEB	3

About Thermal Systems

For each of the normative periods of CTE 2006 [10], CTE 2013 [11] y NZEB, different heat supply systems have been considered; cogeneration, condensing gas boiler and biomass boiler, with the following characteristics (Table **5**):

Table 5: Characteristics heat supply systems

Heat supply	**Nominal Power (kW)**	**Performance (%)**	**Accumulation (l)**	**Solar support (%)**
Cogeneration	1790+109	93/85	10.000	0
Gas boiler	2542	100	4.000	0
Biomass	2542	90		0

A consultation to VISESA, developer of the building, about the most common heating systems of the area has been done to make the proposal of these systems. No complex systems of renewable energy production have been proposed to give a more realistic view to the study.

Combinations of Solutions of Thermal Envelopes and Installations

From all these technical versions, 9 packages with energy solutions have been defined, as shown in Table **6**:

Table 6: Complete relation of the analysed solutions

Combinations assigned	Description
C1	CTE06 + cogeneration
C2	CTE06 + gas boiler
C3	CTE06 + biomass boiler
C5	CTE13 + cogeneration
C6	CTE13 + gas boiler
C7	CTE13 + biomass boiler
C9	NZEB + cogeneration
C10	NZEB + gas boiler
C11	NZEB + biomass boiler

Calculation of Demands and Energy Consumptions

The energy evaluation of the 9 packages of proposed solutions have been done using the informatic tool CERMA [11], authorised by the Ministry of Industry, Energy and Tourism of Spain, as a recognize document to do energy certifications of new and existing residential buildings. The CERMA tool allows the study of an entire building, even complex buildings as the one been studied, with a great variety of heights and orientations. Because it is a tool that does not work with graphical data but with numerical information, an analysis of just one energy model was done. Although, it was necessary to do it like that because the heating system and water heating system is the same for the entire building.

In Table **7** are exemplify the values obtained of energy demand and consumption reached with each one of the packages of solutions, as well as its energy qualification.

Table 7: Demand/consumption of the analysed solutions

			Demand kWh/m^2 year	Consumption kWh/m^2 year	Qualification
Initial			44.10	41.15	A
CTE06	Cogeneration	C1	56.10	51.90	B
CTE06	Gas boiler	C2	56.10	52.40	B
CTE06	Biomass boiler	C3	56.10	80.80	B

Table 7: contd…

CTE13	Cogeneration	C5	43.20	40.32	A
CTE13	Gas boiler	C6	43.20	40.10	A
CTE13	Biomass boiler	C7	43.20	63.20	A
NZEB	Cogeneration	C9	28.10	26.82	A
NZEB	Gas boiler	C10	28.10	26.10	A
NZEB	Biomass boiler	C11	28.10	42.10	A

EVALUATION OF GLOBAL COSTS OF EACH SOLUTION

The costs considered in this study are detailed on the following lines, according to the optimal cost methodology.

Construction Costs of Each Version of the Thermal Envelope

The costs related to the different technical versions of the thermal envelope of the building have been calculated based on the Construction Data Base of the Valencian Community 2014 [12].

From the exigencies require of thermal transmission (U) of each one of the construction elements (façade, roof and empty spaces or gaps) for the comply of the three different normative periods considered (CTE06, CTE13, NZEB), the following costs have been estimated, shown in Table **8**:

Table 8: Relation of construction costs by variant in the thermal envelope

Construction element	**U** (W/m^2K)	**Cost** (€/m^2)	**Cost** (€/ud)
Façade isolation			
CTE06	0,66	54,32	
CTE13	0,27	69,82	
NZEB	0,20	82,02	
Roof isolation			
CTE06	0,21/0,41	45,53/ 74,66	
CTE13	0,21/0,41	45,53/ 74,66	
NZEB	0,21	45,53	
Windows isolation			
CTE06	4,00		507,90 [1]
CTE13	4,00		507,90 [1]

Table 8: contd…

NZEB	1,80		380,93 [1]
Glass isolation			
CTE06	2,70	36,13	
CTE13	2,00	45,63	
NZEB	1,40	73,45	

[1] Average cost taking into account its life cycle (replacement after 20 years)

Maintenance Costs

The maintenance costs have been considered, from the different construction elements as well as the systems. In this sense, it should be highlighted the difficulty to obtain official prices of each of the maintenance operations needed. An approximation to the real value was done, based on a percentage of the construction cost of the elements and systems, variable depending on the major or minor durability of the elements. To do so the maintenance operations of the building have been consulted as they are describe on the Guide of the Building Book [14] that considers the number of replacement of each element or component depending on its life cycle. Afterwards, the derived maintenance costs were estimated using the 2014 Construction Data Base of the Valencian Community, from the Valencian Institute of Building [13].

Energy Costs

The energy prices have been determined from December 2012 and its update is done following the instructions of the European Commission web about energy prices [15].

The price of the biomass is kept constant as it is understood from the analysis of the evolution of prices from the document "Study of an optimal use of renewable energies on the construction sector to be added to the Technical Code of Construction CTE" done by CENER for IDAE [16].

Service Life of the Construction Elements

In Table **9**, it is shown the service life of the construction elements and systems used for the analysis of the optimal cost.

Table 9: Service life of construction elements

Elements	Service life (years)
Thermal envelope and insulation	50 years
Windows	20 years
Systems. Individual equipment	Followed the specifications of Annex A UNE-EN 15459 [17]

Other Needed Data

Other entry relevant data that has been taken into consideration for the calculation of the optimal cost are summarized below:

- A period of calculation is established following the Delegate Regulation [3]. Being a residential building, a period of 30 years is selected for calculation. In general terms, should be taken into account that the impact of the observance time period on the final result is limited due to the greater service life of the construction elements and its residual values.

- In real terms, it is supposed that the price of maintenance and substitution won't rise – in other words, the rise of the nominal price will be equivalent to the general inflation.

- Because of the no inclusion of the professional fees of design and construction supervision on the construction cost of Table **8**, an additional average cost of 10% have been considered over the global cost of construction.

- No other costs have been taken into consideration such as insurances of construction faults, because of its small weight on the total amount.

RESULTS

On the next two graphics (Figs. **5** and **6**) the results of the calculation done for the 9 packages of technical solutions used are presented. The numbers show the difference in costs, in comparison with the requirements of each solution.

- Demand graphic (Fig. **5**): Represents energy demand (Kwh/m^2) of each one of the 9 technical solutions, related to the global costs

(€/m^2). The solutions are represented grouped by colour for each one of the defined heat supplies (cogeneration, gas boiler and biomass boiler). Furthermore, each normative period is defined by figures (square_NZEB, triangle_CTE13, circle_CTE06).

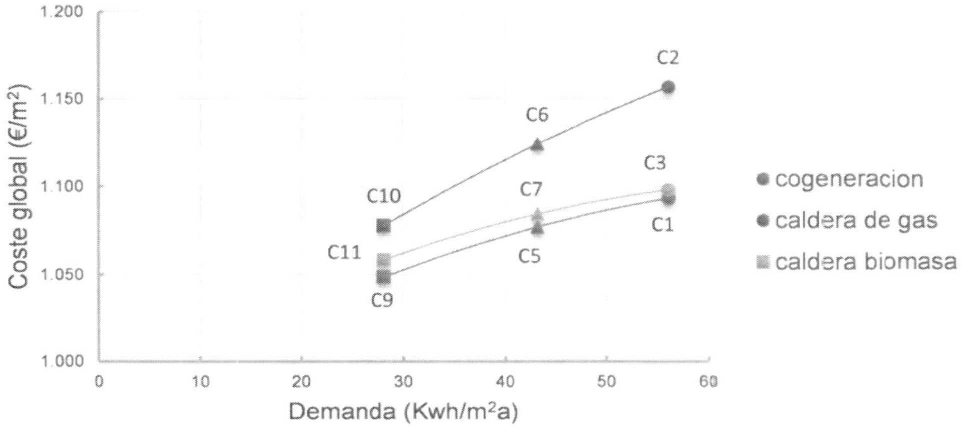

Figure 5: Optimal cost. Demand. (Source: own elaboration).

- Demand graphic (Fig. **6**): Represents the primary energy (Kwh/m^2) of each one of the 9 technical solutions, related to the global costs (€/m^2). The solutions are represented grouped by colour for each one of the defined heat supplies (cogeneration, gas boiler and biomass boiler). Furthermore, each normative period is defined by figures (square_NZEB, triangle_CTE13, circle_CTE06).

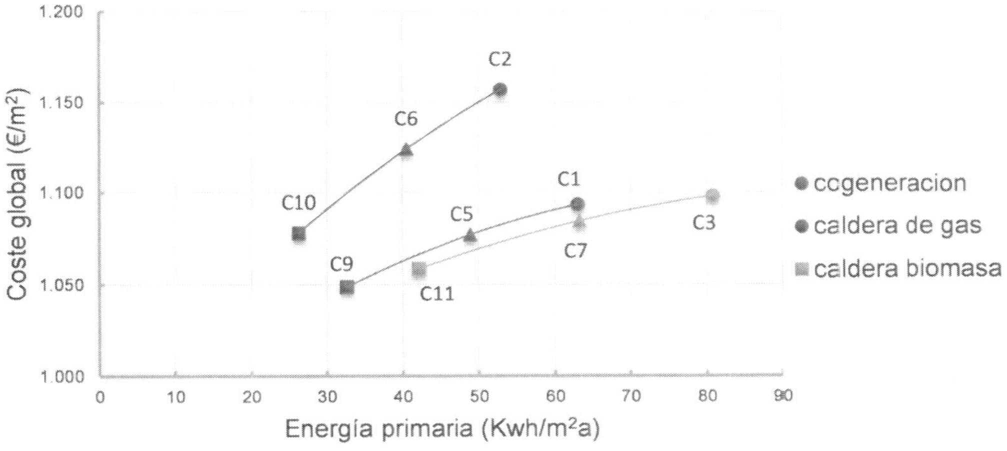

Figure 6: Optimal cost. Primary energy. (Source: own elaboration).

In first place and in a general level, the graphics show that the curves for the different variables of the thermal envelope, with the same heat supply, shows a rising tendency, from NZEB (C9-C10-C11) solutions to the related to CTE06 (C1-C2-C3). For example, on the curve of costs for the variable biomass boiler, even though the tendency is rising, the change of price is just 40€/m^2 over the entire 30 years period of calculation. This represents only 11 cents/m^2 a month. Referred to the variable of heat supply, a major difference between the CTE06 (C1-C2-C3) solutions can be observed, lightly smaller for the related to CTE13 (C5-C6-C7) and a very small differences if we are talking about the NZEB, in values of demand as well as primary energy. Only when we compare solutions introducing variables of heat supply, a greater difference of costs happens. For example, comparing the C1 (CTE06-cogeneration) solutions and C2 (CTE06-gas boiler), with the same thermal envelope solution, the variation of cost is about 231€/m^2 over the entire 30 years period of calculation. This represents 65 cents/m^2 a month. In other words, on the analysis of costs, until reaching the optimal value, the calculation is more sensible for the selection of the heating supply system than for the improvement of the thermal envelope.

As the cogeneration system originally designed for the building shows, the global costs of it are lower in relation with the rest of the system. In the contrary, the most expensive turns to be the gas boiler, a justified result due to the pricing forecast, that is higher than in other types of energy. The biomass system draws our attention. It generates the greatest consumption rates although at a cost level close to the cogeneration system.

Comparing all the technical variables of the thermal envelope as well as of the supply system of heat/cold, the global optimal cost corresponds to the solution of NZEB_Cogeneration, as it is shown at the Table **10**:

Table 10: Optimal global cost

Supplies	**Normative period**		
	CTE06	**CTE13**	**NZEB**
Gas boiler	C2 / 1157	C6 / 1124	C10 / 1078
Biomass boiler	C3 / 1098	C7 / 1084	C11 / 1058
Cogeneration	C1 / 1093	C5 / 1077	C9 / 1049

For each one of the supply systems of heat/cold, its optimal cost is shown in Table **11**. On it can be seen that the optimal cost is mainly corresponded with the NZEB solutions, being common on all the graphics that at a lower cost, a higher reduction of energy. The cost is lower than the energy reduction obtained.

Table 11: Optimal cost for the supply system of heat/cold

Supplies	**Periods**		
	CTE06	**CTE13**	**NZEB**
Gas boiler	C2 / 1157	C6 / 1124	C10 / 1078
Biomass boiler	C3 / 1098	C7 / 1084	C11 / 1058
Cogeneration	C1 / 1093	C5 / 1077	C9 / 1049

CONCLUSION

As a general conclusion, it has been confirmed the difficulty of appliance of the methodology of the optimal cost because some of the data about costs, needed to apply this methodology, are difficult to quantify as it is the case of the maintenance costs. Due to the inexistence of real data, an approximation has been done based on the number of replacements of the elements and its economic costs.

It has been verified that on the evaluation of the optimal cost the parameters related to the heating supply system are more sensible than those related to the improvement of the thermal envelope.

The graphics of optimal cost, about demand as well as consumptions, show that the curves for the variables of the thermal envelope, with a similar heat supply system, have a rising tendency, from the NZEB (C9-C10-C11) solutions to those correspondent to the CTE06 (C1-C2-C3), so the optimal cost will be closer to the requirement established by the NZEB. Regarding the cogeneration case, it is proved that if the building would have been designed with the regulation existing during its construction, CTE2006, the primary no renewable energy consumption would have been of 62 kWh/m^{2}a and the scenario achieved by the NZEB criteria is much lower, 32 kWh/m^{2}a. This shows that if the study would have been done without the optimal cost methodology, the final consumption values would have been similar, but the total cost would have been much higher on the NZEB scenario and, in consequence, it would not have been optimal. Studies done about this subject come to similar conclusions [18, 19].

Referred to the heat supply variables, a greater difference can be seen on the value of optimal cost between the CTE06 (C1-C2-C3) solutions, a little lower for the correspondent to CTE13 (C5-C6-C7) and very small differences of the NZEB cases, on the demand as well as the primary energy values.

The most expensive solution is the gas boiler, because of the pricing forecast, that is higher than in other types of energy. The most efficient solution following the optimal cost methodology is the NZEB, with a system of heat production by cogeneration.

Finally, it is concluded that a building designed following long term economic profitability criteria will be always more energy efficient that if it is only designed following the Construction Code actual requirements. The knowledge generated can be useful to define the NZEB concept and can have very important political implications, because it challenges the idea that high energy efficiency regulations are prohibitive due to the elevated cost that it requires.

ACKNOWLEDGEMENTS

We thank VISESA, public development society of the Basque government, responsible for the construction of the Salburua building. In particular to Alberto Ortiz for all the information facilitated for the elaboration of this study.

CONFLICT OF INTEREST

The authors confirm that this chapter contents have no conflict of interest.

REFERENCES

[1] "Power House Nearly Zero Energy Challenge," 2012. [Online]. Available: http://www.powerhouseeurope.eu/. [Accessed 2014].

[2] European Union, «Directive 2010/31/EU of the European Parliament and of the Council of 19 May 2010 on the energy performance of building,» Official Journal of the European Union, vol. 53, nº 153, pp. 13-35, 2010.

[3] European Union, «Commission Delegated Regulation (EU) No 244/2012 of 16 January 2012 supplementing Directive 2010/31/EU of the European Parliament and of the Council on the energy performance of buildings by establishing a comparative methodology,» *Official Journal of the European Union,* vol. 55, nº 81, pp. 18-36, 2012.

[4] HOUSING EUROPE, "Housing Europe is the European Federation of Public, Cooperative & Social Housing," [Online]. Available: http://www.housingeurope.eu/. [Accessed 2014].

[5] AVS, "The Spanish Social Housing Association," [Online]. Available: http://www.promotores-publicos.org/. [Accessed 2015].

[6] FEDERCASA, "Federazione italiana per la casa", 2002. [Online]. Available: http://www.federcasa.it/. [Accessed 2014].

[7] P. Chalmers, "*Climate Change: Implications for Buildings. Key Findings from theIntergovernmental Panel on Climate Change Fifth Assessment*". Brussels: Buildings Performance Institute Europe, 2014.

[8] BPIE, "*Implementing the cost-optimal methodology in EU countries*". Brussels: Buildings performance Institute Europe, 2013.

[9] C. Charlot-Valdieu y P. Outrequin, “An approach and a tool for setting sustainable energy retrofitting strategies referring to the 2010 EPBD”, *Informes de la Construcción, vol. 63, no EXTRA*, pp. 51-60, 2011.

[10] Ministry of Housing, Spain, “*Royal Decree 314/2006, 17 March 2006 Technical Building Code.Basic Energy Saving Document*”, Official state bulletin Spain, vol. PART II ANNEX, no. 74, 2006.

[11] Ministry of Public Works and Transport Spain, “*Order FOM/1635/2013, 10th September 2013, is the update of the basic document DB-I «Energy saving» of the Spanish Technical Building Code approved by Royal Decree 314/2006, 17th March 2006*”, Official state bulletin Spain, no. 219, pp. 67137-67209, 2013.

[12] FREDSOL Dept. of Applied Thermodynamics UPV, “*CERMA- Abridged method for certification of new and existing residential buildings in SPAIN*”, ATECYR&IVE, 2013.

[13] Valencia Institute of Building (IVE), “*Base de Datos de la Construcción de la Comunidad Valenciana 2014*”. Valencia: IVE, 2014.

[14] B. Serrano-Lanzarote, A. García-Prieto Ruiz y L. Ortega-Madrigal, “*Guía para la elaboración del Libro del edificio*”. Valencia: IVE, 2013.

[15] European Union, "Communication from the commission to the european parliament, the council, the european economic and social committee and the committee of the regions Energy Roadmap 2050," *Official Journal of the European Union, vol. DC0885*, no. 52011, 2011.

[16] CENER, “*Estudio de aprovechamiento óptimo de las energías renovables en la edificación para su incorporación al Código Técnico de la Edificación CTE*”. Madrid: IDAE, 2011.

[17] AENOR-CEN, UNE-EN 15459:2008, “*Eficiencia energética de los edificios. Procedimiento de evaluación económica de los sistemas energéticos de los edificios*”. Asociación Española de Normalización (AENOR), 2008.

[18] J. Kurnitskia, A. Saarib, T. Kalameesc, M. Vuolled, J. Niemeläd y T. Tarke, “Cost optimal and nearly zero (nZEB) energy performance calculations for residential buildings with REHVA definition for nZEB national implementation”, *Energy and Buildings*, vol. 43, no 11, p. 3279-3288, 2011.

[19] J. Morrissey y R. Horne, “Life cycle cost implications of energy efficiency measures in new residential buildings”, *Energy and Buildings*, vol. 43, no 4, pp. 915-924, 2011.

Chapter 4: Development of an Assessment Tool for Building Envelope Retrofit Based on Environmental Indicator

CHAPTER 4

Development of an Assessment Tool for Building Envelope Retrofit Based on Environmental Indicator

Patricia Huedo[1], Belinda López-Mesa[2,*] and Elena Mulet[1]

[1]*Universitat Jaume I, Spain and* [2]*Universitat de Zaragoza, Spain*

Abstract: We are witnessing a profound change towards a new and necessary relationship with the building stock in Spain, which needs to change from a repairing-based culture to a building sustainability management culture, by which active prevention, conservation and maintenance increase the quality of life of the occupants thereof and reduce use and maintenance costs. To achieve the foregoing goals, new reliable tools are required which shall provide values for significant indicators in relation to building sustainability. The main aim of this paper is to define a model to evaluate the environmental impacts associated to the energy retrofit of building envelopes in order to assign them a value by means of recognized sustainability indicators. Indeed, this evaluation model aims to contribute to the development of a tool to support designers, which shall allow them to compare the environmental behavior of different alternatives for the envelope retrofit. The proposed methodology is based on the application of a simplified LCA. The defined evaluation model provides values for the following indicators: CO_2 emissions; energy consumption; consumption of potable water; hazardous and non-hazardous waste; and maintenance costs. It makes the designer's tasks of comparison and selection easier by means of the characterization of building envelope retrofit alternatives which combine different materials which are involved in construction assemblies and solutions. It also provides a comparative scale that allows the deployment of the impacts of each of the life cycle phases of the building which provides for the easy detection of the most critical phase or impact thereof.

Keywords: Building energy retrofit, building envelope, building sustainability management, designer support, durability, existing buildings, evaluation model, Life Cycle Analysis, maintenance costs, refurbishment, sustainability indicators.

INTRODUCTION

Since 1997, being the year in which the Kyoto Protocol was signed within the United Nations Framework Convention on Climate Change (UNFCCC), environmental protection has become a global need and a growing urgency exists

***Corresponding author Belinda López-Mesa:** Predepartment of Architecture, University of Zaragoza, Spain; E-mail: belinda@unizar.es

Pilar Mercader-Moyano (Ed.)

for the incorporation of the sustainability principles into all of our productive processes.

In fulfillment of the "20/20/20" objective proposed by the European Union for 2020, the Directive 2010/31/EU promotes the reduction of energy consumption in the built environment, emphasizing that the construction sector is one of the main sources of CO_2 emissions. This Directive promotes the establishment of requirements regarding profitability or cost optimization guidelines that stimulate the use of certain technical assemblies or design solutions that provide for the better energy efficiency of buildings. Likewise, the European Directives 2012/27/EU4 and 2013/12/EU5 establish the obligation to retrofit buildings which belong to Public Administrations that do not meet the minimum parameters established for energy consumption and CO_2 emissions, while maintaining the required comfort level thereof. Therefore, it is essential that the entire building stock becomes sustainable and accordingly, in order to achieve this, it is necessary to carry out the energy retrofit of all buildings the level of energy demand of which exceeds the permitted maximums.

We understand that this is the beginning of a profound change towards a new and necessary relationship with the building stock in Spain, which needs to change from a repairing-based culture to a building sustainability management culture, by which active prevention, conservation and maintenance increase the quality of life of the occupants thereof and reduce use and maintenance costs. To achieve the foregoing goals, new reliable tools, that is to say, scientifically supported, are required which shall provide values for significant indicators in relation to building sustainability.

All the options regarding energy consumption reduction for buildings inexcusably require a reduction of energy demands, and that is why the design decisions need at their disposal valid technological alternatives as well as suitable and objective information regarding said options, together with instruments that enable the integral evaluation thereof [1]. It is important to emphasize the importance of the study of the environmental impacts produced by the use of certain construction materials, but it is also important to emphasize the impact of the combination of all said materials when they are used in a specific construction system/assembly [2]. Thus, the designer faces the difficulty of combining two different but closely-linked scenarios; he/she needs to know to what extent the materials or the construction solutions used for the building retrofit are respectful to the environment and at the same time to what extent they are capable of guaranteeing a reduction of the energy demand of the refurbished building, maintaining the conditions of comfort of the building throughout the residual life thereof. One of the key aspects in any construction process is the design of the building envelope, and even more so taking into account that the design decisions have a significant impact on the project energy environmental context. For this reason, the instruments used to evaluate the envelope must consider the energy and sustainability aspects in an integral way [3]. Throughout the world, an extensive variety of tools based on the LCA methodology have been specifically developed to support designers during the initial phase of design, which evaluate the building components taking into consideration all of the phases of the life cycle [4, 5].

AIM AND RESEARCH METHODOLOGY

The main purpose of this paper is to define an environmental evaluation model for different construction assemblies and solutions for carrying out energy retrofits of the envelope of existing buildings, which allows to value the impact generated during the refurbishment process as well as the impacts associated with the envelope during the use phase of the building after the retrofit, by means of the selection of certain materials or construction assemblies.

The proposed methodology represents the development of the second version of a model previously developed for new buildings for the evaluation of envelope construction solutions based on environmental indicators, which consisted of the application of a simplified LCA to buildings taking into consideration different combinations of construction assemblies [5].

The difference between both models is that the first model was created during the repairing culture, in which for the use phase only reparations of the envelope materials were considered. The new model is defined in the new era of smart building management, and consequently it produces indicators for the eventual envelope energy retrofits during the use phase.

The methodology that was followed to develop this model is summarized in the diagram below (Fig. **1**).

DEVELOPMENT OF THE VERSION OF AN ASSESSMENT MODEL TO EVALUATE THE ENVIRONMENTAL BEHAVIOR OF THE ENVELOPE RETROFIT CONSTRUCTION SOLUTIONS

Background

The first version of the developed model calculates the environmental impacts in relation to the materials during the phase of manufacture and installation, as well as the environmental impacts associated with the envelope, related to CO_2 emissions and to energy consumption during the use phase of the building, due to the air conditioning installation. Likewise, the model provides information regarding the impacts produced by water consumption during the manufacturing and execution phases of the building and regarding the wastes generated together with an economic valuation of the building during its life cycle.

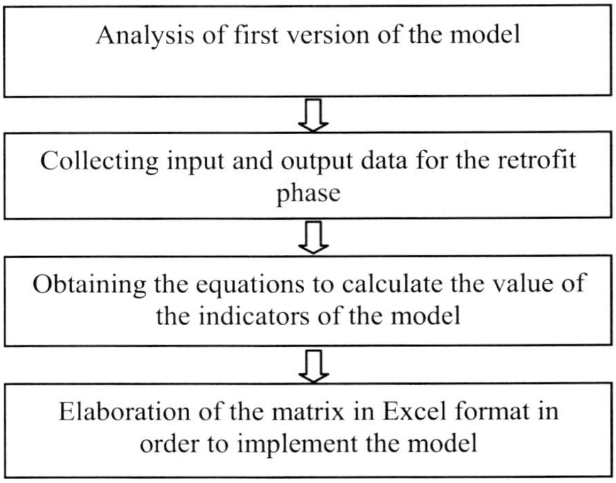

Figure 1: Methodological overview of the research.

To adequately value the impacts generated by the envelope of buildings, a state of the art study was conducted [1, 5], the aim of which was to analyze other studies that proposed different methodologies for the evaluation of the environmental impact of the envelope of buildings, taking into consideration both the impacts of materials and assemblies as well as the impact regarding heating and air conditioning [6-12]. All these studies conclude that it is necessary to link the demand of the building to the construction solutions of the envelope using energy simulation software or through the real measurement of its performance.

The methodology adopted for the first version of the model was based on obtaining the necessary information to develop the equations of the sustainability indicators previously selected in each of the phases of the life cycle of the building. To assess the foregoing, the construction assemblies commonly used for building in Spain were evaluated.

Every construction solution was designated by a capital letter that indicated the class (C = roof, F = façade and H = carpentry) and a numerical index that indicated the type of construction assembly pursuant to the following description:

- C1 Flat roof, conventional, non-ventilated, trafficable, with fixed flooring.

- C2 Flat roof ventilated, trafficable, with fixed flooring.

- C3 Flat roof with insulation, conventional, non-trafficable.

- F1 Brick cavity walls, with outer wall of facing bricks, 5 cm thick insulation.

- F2 Brick cavity walls, with coated outer masonry wall, 5 cm thick insulation.

- F3 Brick cavity walls, with coated outer masonry wall, 10 cm thick insulation.

- F4 Back-ventilated façade of masonry wall, 5 cm thick insulation.

- F5 curtain wall.

- H1 Aluminium windows with thermal bridge breaking system.

- H2 PVC windows.

- H3 Wood windows.

The different combinations of the construction elements resulted in 45 building assemblies for the envelope which are designated by means of the three letters with their corresponding indices. Calculations for the use phase were then carried out for the 45 combinations of the building envelope, in two different climate zones, B3 and E1. Climate zone B3 constitutes a temperate humid climate and climate zone E1 constitutes a cold dry climate. Furthermore, calculations were performed in two building orientations, North-east orientation (NE) and South-east orientation (SE). Altogether 45 combinations were evaluated in two orientations and two climate zones, that is to say a total of 180 options. All the values were obtained for the particular case of a two-story semi-detached house with a floor area of 93.5 m^2, the functional unit being taken as 1 m^2 of usable floor area and with a life span of 50 years.

To analyze the impacts of the manufacturing, installation and maintenance phases of the building, we selected a tool with a database with environmental information of Spanish construction materials and products. The foregoing tool was TCQ200 and its environmental management module TCQ GMA [13]. This software, developed by the *Instituto de Tecnología de la Construcción de Cataluña* (IteC), uses as a database the bank BEDEC, PR/PCT, which has been created thereby [14], which integrates costs and environmental data. The environmental data is based on the analysis of different European databases and not on audits. This could explain certain differences in relation to specific values in comparison to other tools. On the other hand, it is easy access software in light of the fact that it is offered to researchers and students at a low price and it has been used in numerous research studies, such as in [7, 8, 12].

The steps followed for obtaining the values in the different phases of the life cycle of the building are the following:

- In the manufacturing and installation phases the results of the impacts obtained by the tool TCQGMA produced by the different construction assemblies (roofs, façades and carpentries) were used, thereby obtaining the equivalent emission of CO_2, consumptions of primary energy, consumptions of water, hazardous and non-hazardous wastes and investment costs. In this phase of the calculation, the units were expressed by square meter of surface area of the built assembly.

- In the maintenance phase, the impacts produced by the different construction assemblies (roofs, façades and carpentries) were obtained by introducing the materials of each construction solution capable of being replaced throughout 50 years into the tool TCQGMA, in order to obtain the CO_2 emissions, the energy consumptions and the maintenance costs, depending upon the proportion of material that integrates each construction solution. In this phase of the calculation, the units were expressed by square meter of surface area of the built assembly.

- During the use phase, consumptions and CO_2 emissions, resulting from heating and air conditioning systems, were calculated using the energy simulation tools LIDER [15] and CALENER VYP [16]. The functional unit was 1 m^2 of usable floor area.

From the results with the aforementioned steps, equations were obtained for the different indicators in each analyzed phase as set out herein below.

In the manufacturing, installation and maintenance phases, the indicators were obtained by calculating a common factor for each construction assembly group and for each impact that corresponds to the lowest value of the impact in each group. The factors or coefficients of each impact were then obtained by dividing their real value by the lowest impact value for each group. FS indicates the surface factor that represents the proportion of a construction assembly per m^2 of usable floor area of the building.

The equations for the use phase were obtained by producing a multiple linear regression model, using the analyzed factors in the case under study, *i.e.* climate zone, building orientation, three types of roofing, five types of façade and three types of carpentry, as the explanatory variables, whereby the impacts during the use phase were taken as dependent variables.

The equation for each indicator was obtained by selecting the data from the column for the corresponding indicator (dependent variable) that is to be calculated and the data corresponding to the columns for all the explanatory variables. The regression model was used to obtain the value of the coefficients for each of the dependent variables. The independent term of each equation was established by the regression model, and the coefficient of each variable was replaced in each case according to the roofing, façade and carpentry option chosen by the planner and depending on the building orientation and the climate zone.

To apply the model to other building types, in its first version, we calculated a correction factor (F_tviv) according to the type of housing construction. This was achieved by comparing the resulting impacts in the use phase of the two-story semi-detached house (used as reference) with those of a residential block and a detached family house. A measure of proportionality was noted. To obtain the correction factor, the mean value of the distance between each of the impacts during the use phase of the residential block and the detached family house and those of the semi-detached house was calculated, taking each climate zone into consideration separately.

The equations obtained in the first version of the model were implemented in a number of matrices in Excel. This allows different alternative designs to be chosen for the building envelope in different climate zones and orientations in three types of buildings, so that the estimated values of the impacts generated in each of the phases of the life cycle of the building can be displayed automatically (Fig. **2**).

Data Collection to Complete the Inventory Analysis of the Retrofit Phase

For the development of the second version of the model, information has been obtained on the environmental behavior of different construction assemblies, also using a simplified life cycle analysis (LCA) methodology. This model aims to value the impacts of the building from the moment when the refurbishment process begins in order to support the adoption of solutions with low environmental impacts, extended longevity and the ability to ensure an optimal energy efficiency improvement for the building. For this reason, the model does not consider the impacts that were generated in the original construction phase of the building, which could be assessed with the first version of the model.

To achieve this, the construction solutions which are commonly used in Spain have been evaluated taking into consideration solely and exclusively the materials which constitute said assemblies that can be replaced, repaired or improved throughout the useful life of the building based upon a functional unit.

The functional unit defines the function of a system of the product, which is used as a reference unit in a LCA study. In this case, 1 m^2 of usable floor area has been selected as the functional unit, considering that the aim of different construction assemblies, in order to ensure the comfort conditions, is to work as an envelope to ensure a certain level of comfort in a specific usable floor area.

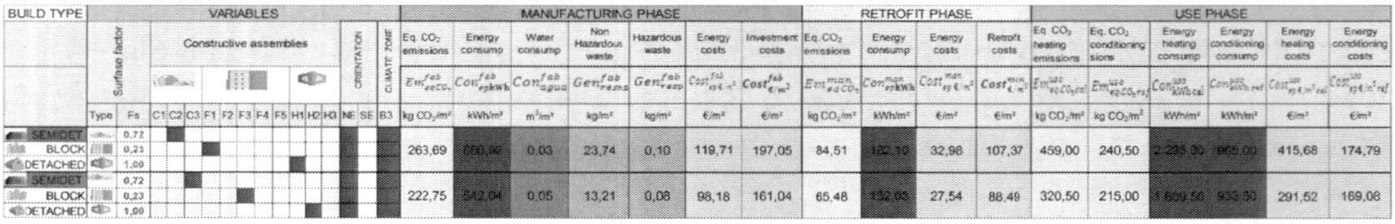

BUILD TYPE	Surfase factor Fs	Constructive assemblies (C1 C2 C3 F1 F2 F3 F4 F5 H1 H2 H3)	ORIENTATION (NE SE)	CLIMATE ZONE	MANUFACTURING PHASE: Eq. CO_2 emissions (kg CO_2/m²)	Energy consump (kWh/m²)	Water consump (m³/m²)	Non Hazardous waste (kg/m²)	Hazardous waste (kg/m²)	Energy costs (€/m²)	Investment costs (€/m²)	RETROFIT PHASE: Eq. CO_2 emissions (kg CO_2/m²)	Energy consump (kWh/m²)	Energy costs (€/m²)	Retrofit costs (€/m²)	USE PHASE: Eq. CO_2 heating emissions (kg CO_2/m²)	Eq. CO_2 conditioning sions (kg CO_2/m²)	Energy heating consump (kWh/m²)	Energy conditioning consump (kWh/m²)	Energy heating costs (€/m²)	Energy conditioning costs (€/m²)
SEMIDET	0,72	C2	NE	B3																	
BLOCK	0,23	F1	NE	B3	263,69	660,00	0,03	23,74	0,10	119,71	197,05	84,51	182,10	32,98	107,37	459,00	240,50	2 295,00	865,00	415,68	174,79
DETACHED	1,00	H1	NE	B3																	
SEMIDET	0,72	C3	NE	B3																	
BLOCK	0,23	F3	NE	B3	222,75	542,04	0,05	13,21	0,08	98,18	161,04	65,48	132,05	27,54	88,49	320,50	215,00	1 609,50	933,50	291,52	169,08
DETACHED	1,00	H2	NE	B3																	

Figure 2: Calculation matrix: result of the estimated values of each impact in the different phases of the life cycle.

In relation to the life span period, and bearing in mind that the building to be retrofitted has already consumed a part of its life cycle, we will refer to it as the period of residual life from the moment that the building becomes operational after its refurbishment. Buildings have a life expectancy of between 50 and 100 years. To carry out this research, we will consider a total life span of 50 years.

The designation of the construction assemblies is the same as that used in the first version: every construction assembly is designated by a capital letter indicating the class (C = roof, F = façade and H = carpentry) and a numerical index that indicates the type.

To carry out the evaluation of the impacts corresponding to the refurbishment phase, each of the materials which constitute each construction assembly are assigned a reconditioning factor F_R depending on the number of times that the above mentioned material has to be replaced throughout the total life span of the building [17].

$$F_R = V_U/dur \quad (1)$$

Where F_R= Reconditioning factor

V_U= Building life span in years

dur= Durability in years

For example, a reconditioning factor of 5 has been assigned to constituting materials that may be replaced 5 times throughout the useful life of the building.

To establish the durability of every element within the construction assemblies, the recommendations of the ISO Norm 15686-1 have been followed. According to the foregoing standard, the useful life of each construction assembly is obtained by taking into consideration an average between the minimum and maximum durability of every component. The durability values of the materials which comprise each assembly used to calculate their F_R have been obtained by checking different sources and databases [18, 19].

The assessment of the impacts during the refurbishment phase is carried out by using the results obtained by the TCQGMA tool of the TCQ2000 program. The values are obtained by multiplying the impact for the reconditioning factor F_R minus 1 to discount the impact during the manufacturing and installation phases.

The final impact imputable to the maintenance and use phase after the building retrofit considering exclusively the period of residual life are obtained by multiplying the result by the residual life span of the building divided by the total life span, as described in the following equation.

$$I_{Vures}=I_{Vu}\ V_{Ures}/V_U \quad (2)$$

Where

I_{Vures} = Impact attributable to the maintenance and use phase after the building retrofit

I_{Vu} = Impact attributable to the maintenance and use phase before and after the building retrofit

V_{Ures} = Residual life span in years

V_U= Total life span, in years

On the other hand, considering that a great part of the buildings to retrofit lack suitable thermal insulation, the impact generated by the envelope thermal insulation is evaluated without discounting the initial impact; that is to say, we have considered that the thermal insulation shall be added after having carried out the envelope energy retrofit. Moreover, the impact generated by the substitution of the carpentries are valued for all of the different cases, taking into consideration that the carpentries of a significant part of the buildings to be refurbished do not conform with the minimum transmittances required pursuant to Spanish Regulations [20].

Tables **1**, **2** and **3** show the construction assemblies that are most commonly used for the building envelope and accordingly which were chosen for this study as well as the specific parameters adopted to complete an inventory of the refurbishment phase. The first row indicates the construction assembly designated by a capital letter and a numerical index. For example C1 is the code assigned to identify a flat roof, conventional, non-ventilated and trafficable; F1 is the code for brick cavity walls, with outer wall of facing bricks, and 5 cm thick insulation; and H1 is the code for aluminium windows with thermal bridge breaking system. The following four columns show the materials which constitute said construction assemblies as well as their thickness, durability and their reconditioning factor F_R in the corresponding units, all of which are expressed by 1 m^2 of built assembly. The next three columns show the values corresponding to the manufacturing and installation phases and the last three columns on the right contain the values corresponding to the refurbishment phase or corrective maintenance during the entire life span of the building.

In Tables **1**, **2** and **3**, the materials that will necessarily be added or replaced during the energy retrofit are shaded. All the values are expressed by m^2 of built assembly, taking into consideration a life span of the building of 50 years.

Table 1: Impacts due to the roof materials, during the manufacturing, installation and retrofit phases

C1	Flat roof, conventional, non-ventilated, trafficable				MANUFACTURING AND INSTALLATION PHASE			RETROFIT PHASE		
	Material	Thickness m	Durability years	F_R	Investment cost €/m^2	Emissions kgCO_2/m^2	Energy consumption kWh/m^2	Emissions · F_R kgCO_2/m^2	Energy consumption · F_R kWh/m^2	Investment Costs· F_R €kWh/m^2
1	**P_ Finishing ceramic tiles**	0.010	50	1.00	36.44	41.67	103.16	0.00	0.00	0.00
2	**MA_ Mortar**	0.040	37	1.35	6.91	7.02	10.51	3.69	2.47	2.43
3	**Csa_ Geotextile, polypropylene 125 gr/m^2**	0.00013	30	1.67	2.14	1.67	3.15	2.10	1.11	1.43
4	**I_ Waterproof sheet bitumen LO-40/FV**	0.007	23	2.17	33.12	34.69	67.57	79.32	40.72	38.88
5	**Cs_ Geotextile, polypropylene 125 gr/m^2**	0.00013	30	1.67	2.14	1.67	3.15	2.10	1.11	1.43
6	**AT_ Thermal insulation XPS**	0.050	60	0.83				27.20	44.65	8.20
7	**B_ Vapour barrier**	0.005	61	0.82	4.62	7.13	13.48	-2.43	-1.29	-0.83
8	**FP_ Aerated concrete for roof slope**	0.048	50	1.00	10.16	44.05	61.35	0.00	0.00	0.00
9	**SR_ Reinforced concrete one-way slab, ceramic hollow plot**	0.300	50	1.00	134.92	472.28	948.13	0.00	0.00	0.00
10	**RI_ Plastering**	0.010	46	1.09	4.83	1.7	5.33	0.46	0.15	0.42
Total		0.470			235.28	611.88	1215.83	112.45	88.93	51.95
C2	Flat roof ventilated, trafficable				MANUFACTURING AND INSTALLATION PHASE			RETROFIT PHASE		
	Material	Thickness m	Durability years	F_R	Investment cost €/m^2	Emissions kgCO_2/m^2	Energy consumption kWh/m^2	Emissions F_R kgCO_2/m^2	Energy consumption · F_R Wh/m^2	Investment Costs· F_R €kWh/m^2
1	**P_ Finishing ceramic tiles**	0.010	50	1.00	36.44	41.67	103.16	0.00	0.00	0.00
2	**MA_ Mortar**	0.040	37	1.35	6.91	7.02	10.51	2.47	3.69	2.43
3	**Csa_ Geotextile, polypropylene 125 gr/m^2**	0.00013	30	1.67	2.14	1.67	3.15	1.11	2.10	1.43
4	**I_ Double waterproof sheet bitumen LO-40/FV**	0.007	23	2.17	33.12	34.69	67.57	40.72	79.32	38.88
5	**Cs_ Geotextile, polypropylene 125 gr/m^2**	0.00013	30	1.67	2.14	1.67	3.15	1.11	2.10	1.43
6	**FP_ Ceramic tiles for roof slope**	0.200	60	0.83	16.96	13.57	6.51	-2.26	-1.09	-2.83
7	**C_ Ventilated air chamber**	0.003						0.00	0.00	0.00
8	**AT_ Thermal Insulation mineral wool [0.04 W/mK]**	0.005	50	1.00				3.04	12.77	6.19

Table 1: contd…

9	**SR_ Reinforced concrete one-way slab, ceramic hollow plot**	0.300	50	1.00	134.92	472.28	948.13	0.00	0.00	0.00
10	**RI_Plastering**	0.010	46	1.09	4.83	1.7	5.33	0.15	0.46	0.42
Total		0.575			237.46	574.27	1147.51	46.34	99.36	47.94
C3	**Flat roof with insulation, conventional, non-trafficable.**				**MANUFACTURING AND INSTALLATION PHASE**			**RETROFIT PHASE**		
	Material	Thickness m	Durability years	F_R	Investment cost €/m^2	Emissions kgCO$_2$/m^2	Energy consumption kWh/m^2	Emissions · F_R kgCO$_2$/m^2	Energy consumption · F_R kWh/m^2	Investment Costs· F_R €kWh/m^2
1	**P_Finishing gravel**	0.020	50	1.00	36.44	41.67	103.16	0.00	0.00	0.00
2	**Csa_ Geotextile, polypropylene 125 gr/m^2**	0.00013	30	1.67	2.14	1.67	3.15	1.11	2.10	1.43
3	**AT_ thermal insulation XPS**	0.050	60	0.83				27.2	44.65	8.20
4	**Cs_ Geotextile, polypropylene 125 gr/m^2**	0.002	30	1.67	33.12	34.69	67.57	23.13	45.05	22.08
5	**I_ Double waterproof sheet bitumen LO-40/FV**	0.007	23	2.17	2.14	1.67	3.15	1.96	3.70	2.51
6	**FP_ Aerated concrete for roof slope**	0.200	50	1.00	10.16	44.05	61.35	0.00	0.00	0.00
7	**SR_ Reinforced concrete one-way slab, ceramic hollow plot**	0.300	50	1.00	134.92	472.28	948.13	0.00	0.00	0.00
8	**RI_ Plastering**	0.010	46	1.09	4.83	1.7	5.33	0.15	0.46	0.42
Total		0.589			223.75	597.73	1191.84	53.55	95.96	34.64

Table 2: Impacts due to the façade materials, during the manufacturing, installation and retrofit phases

F1	**Brick cavity walls, with outer wall of facing bricks, 5 cm thick insulation**				**MANUFACTURING AND INSTALLATION PHASE**			**RETROFIT PHASE**		
	Material	Thickness m	Durability years	F_R	Investment cost €/m^2	Emissions kgCO$_2$/m^2	Energy consumption kWh/m^2	Emissions · F_R kgCO$_2$/m^2	Energy consumption · F_R kWh/m^2	Investment Costs· F_R €kWh/m^2
1	**LC_ Exterior masonry wall of ceramic perforated brick of 12 cm thick, with cement mortar joints**	0.12	81	1.00	52.91	45.64	152.72	0.00	0.00	0.00
2	**RM_ Intermediate coat. A plaster on the interior face of the principal with cement mortar (1:6)**	0.02	37	1.35	12.88	4.1	6.14	1.44	2.16	4.53
3	**C_ Non-ventilated air chamber**	0.05						0.00	0.00	0.00
4	**AT_ Thermal Insulation, mineral wool**	0.05	50	1.00				3.04	12.77	6.19
5	**LH_ Inner skin of double hollow ceramic brick 7cm thick with cement mortar joints (1:6)**	0.07	81	1.00	21.43	12.72	41.23	0.00	0.00	0.00

Table 2: contd…

6	**RI_ Plastering**	0.02	46	1.09	4.83	1.7	5.33	0.15	0.45	0.42
Total		0.315			92.05	64.16	205.42	4.53	15.39	11.14
F2	**Brick cavity walls, with coated outer wall of brickwork, 5 cm thick insulation**				**MANUFACTURING AND INSTALLATION PHASE**			**RETROFIT PHASE**		
	Material	Thickness m	Durability years	F_R	Investment cost €/m^2	Emissions kgCO_2/m^2	Energy consumption kWh/m^2	Emissions · F_R KgCO_2/m^2	Energy consumption · F_R kWh/m^2	Investment Costs· F_R €kWh/m^2
1	**RE_ Continuous outer coating with cement mortar (1:6)**	0.02	37.00	1.35	12.88	4.10	6.14	1.44	2.16	4.53
2	**LC_ Exterior masonry wall of double hollow ceramic brick 11.5 cm thick with cement mortar joints (1:6)**	0.12	81.00	1.00	28.02	21.32	68.97	0.00	0.00	0.00
3	**C_ Non-ventilated air chamber**	0.05						0.00	0.00	0.00
4	**AT_ Thermal Insulation, mineral wool**	0.05	50.00	1.00				3.04	12.77	6.19
5	**LH_ Inner layer of double hollow ceramic brick 7cm thick with cement mortar joints (1:6)**	0.07	81.00	1.00	21.43	12.72	41.23	0.00	0.00	0.00
6	**RI_ Plastering**	0.02	46.00	1.09	4.83	1.70	5.33	0.15	0.46	0.42
Total		0.32			67.16	39.84	121.67	4.63	15.39	11.14
F3	**Brick cavity walls, with coated outer wall of brickwork, 10 cm thick insulation**				**MANUFACTURING AND INSTALLATION PHASE**			**RETROFIT PHASE**		
	Material	Thickness m	Durability years	F_R	Investment cost €/m^2	Emissions kgCO_2/m^2	Energy consumption kWh/m^2	Emissions · F_R kgCO_2/m^2	Energy consumption · F_R kWh/m^2	Investment Costs· F_R €kWh/m^2
1	**RE_ Continuous outer coating with cement mortar (1:6)**	0.02	37.00	1.35	12.88	4.10	6.14	1.44	2.16	4.53
2	**LC_ Exterior masonry wall of double hollow ceramic brick 11.5 cm thick with cement mortar joints (1:6)**	0.12	81.00	1.00	28.02	21.32	68.97	0.00	0.00	0.00
3	**C_ Non-ventilated air chamber**	0.05						0.00	0.00	0.00
4	**AT_ Thermal Insulation, mineral wool**	0.10	50	1.00				21.55	9.62	17.96
5	**LH_ Inner layer of double hollow ceramic brick 7cm thick with cement mortar joints (1:6)**	0.07	81.00	1.00	21.43	12.72	41.23	0.00	0.00	0.00
6	**RI_ Plastering**	0.02	46.00	1.09	4.83	1.70	5.33	0.15	0.46	0.42
Total		0.37			67.16	39.84	121.67	23.14	12.24	22.91
F4	**Back-ventilated façade of brickwork, 5 cm thick insulation**				**MANUFACTURING AND INSTALLATION PHASE**			**RETROFIT PHASE**		
	Material	Thickness m	Durability years	F_R	Investment cost €/m^2	Emissions kgCO_2/m^2	Energy consumption kWh/m^2	Emissions · F_R kgCO_2/m^2	Energy consumption · F_R kWh/m^2	Investment Costs· F_R €kWh/m^2
1	**RE_ Outer discontinuous coating of ceramic tiles mechanically fastened with aluminium substructure type T**	0.03	45	1.11	87.59	17.91	80.6	1.99	8.96	9.73
2	**C_ Ventilated air chamber**	0.05						0.00	0.00	0.00
5	**AT_ Thermal Insulation, mineral wool**	0.10	50	1.00				3.04	12.77	6.19
	RM_ Continuous outer coating with cement mortar (1:6)	0.02	37	1.35	4.83	1.7	5.33	0.60	1.87	1.70

Table 2: contd…

6	**LC_ Inner layer of double hollow ceramic brick 11.5 cm thick with cement mortar joints (1:6)**	0.12	81	1.00	28.02	21.32	68.97	0.00	0.00	0.00
7	**RI_ Plastering**	0.02	46	1.09	4.83	1.7	5.33	0.15	0.46	0.42
Total		0.33			125.27	42.63	160.23	5.78	24.06	18.04

F5	**Curtain Wall**				**MANUFACTURING AND INSTALLATION PHASE**			**RETROFIT PHASE**		
	Material	Thickness m	Durability years	F_R	Investment cost €/m²	Emissions $kgCO_2/m^2$	Energy consumption kWh/m²	Emissions · F_R $KgCO_2/m^2$	Energy consumption · F_R kWh/m²	Investment Costs· F_R €kWh/m²
1	**Double glazing 6/8/6, low E glass**	0.01	40	1.25	122.67	51.07	235.93	12.77	58.98	30.67
2	**Aluminium substructure of tubular mullions and horizontals transoms**		50	1.00	274.57	36.97	179.02	0.00	0.00	0.00
3	**Dry air 8 mm space**	0.01						0.00	0.00	0.00
5	**Aluminium Composite Panel**	0.00	50	1.00	48.06			0.00	0.00	0.00
Total		0.02	50		445.30	88.04	414.95	12.77	58.98	30.67

Table 3: Impacts due to the carpentry materials, during the manufacturing, installation and retrofit phases

H1	**Aluminium windows with thermal bridge breaking system**				**MANUFACTURING AND INSTALLATION PHASE**			**RETROFIT PHASE**		
	Material	Thickness m	Durability years	F_R	Investment cost €/m²	Emissions $kgCO_2/m^2$	Energy consumption kWh/m²	Emissions · F_R $kgCO_2/m^2$	Energy consumption · F_R kWh/m²	Investment Costs· F_R €kWh/m²
1	**Double glazing 6/6/6, low E glass**	0.01	40.00	1.25	122.67	51.07	235.93	63.84	294.91	153.34
2	**Aluminium frame with thermal bridge breaking system**	0.00	50.00	1.00	183.00	592.00	1203.72	592.00	1203.72	183.00
3	**Filler of neutral silicon**	0.005	20.00	2.50	4.76	1.29	2.43	3.23	6.08	11.90
4	**Dry air space 6mm thick**	0.01								
5	**Ironwork of steel**	0.02	50.00	1.00						
6	**Galvanized steel 40x20mm subframe**	0.02	50.00	1.00						
Total		0.05			310.43	644.36	1442.08	659.06	1504.71	348.24

Table 3: contd…

H2	**PVC windows**				**MANUFACTURING AND INSTALLATION PHASE**			**RETROFIT PHASE**		
	Material	Thickness m	Durability years	F_R	Investment cost €/m²	Emissions $kgCO_2/m^2$	Energy consumption kWh/m^2	Emissions F_R $kgCO_2/m^2$	Energy consumption · F_R kWh/m^2	Investment Costs· F_R €kWh/m²
1	**Double glazing 6/6/6, low E glass**	0.01	40	1.25	122.67	51.07	235.93	63.84	294.91	153.34
2	**PVC frame with three chambers**	0.00	60	1.00	171	292.1	597.08	292.10	597.08	171.00
3	**Filler of neutral silicon**	0.005	20	2.50	4.76	1.29	2.43	3.23	6.08	11.90
4	**Dry air space 6mm thick**	0.01								
5	**Ironwork of steel**	0.02	50	1.00						
6	**Galvanized steel 40x20mm subframe**	0.02	50	1.00						
Total		0.02			298.43	344.46	835.44	359.16	898.07	336.24
H3	**Wood windows**				**MANUFACTURING AND INSTALLATION PHASE**			**RETROFIT PHASE**		
	Material	Thickness m	Durability years	F_R	Investment cost €/m²	Emissions $kgCO_2/m^2$	Energy consumption kWh/m^2	Emissions · F_R $kgCO_2/m^2$	Energy consumption · F_R kWh/m^2	Investment Costs· F_R €kWh/m²
1	**Double glazing 6/6/6, low E glass**	0.01	40	1.25	122.67	51.07	235.93	63.84	294.91	153.34
2	**High density wood frame**	0.02	40	1.25	200.11	3.85	20.75	4.81	5.19	250.14
3	**Filler of neutral silicon**	0.005	20	2.50	4.76	1.29	2.43	3.23	3.65	11.90
4	**Dry air space 6mm thick**	0.01								
5	**Ironwork of steel**	0.02	50	1.00						
6	**Wood frame 40x20mm**	0.02	50	1.00						
Total		0.04	30		327.54	56.21	259.11	71.88	303.75	415.38

The rest of values obtained for the inventory analysis of all the evaluated construction assemblies are synthesized in Table **4**, where only results are shown.

Obtaining the Equations for Calculating the Impacts During the Energy Retrofit Phase

The equations for the indicators during the energy refurbishment phase were obtained by using the results of the impacts obtained by the TCQGMA tool displayed in Tables **1**, **2** and **3**. The valuation of the costs of the energy consumption has been carried out by multiplying the price of the kW·h (€0.181126) at the moment in which the research was carried out by the consumption in kW·h due to the manufacturing and installation of the materials in the envelope retrofit process.

Table **4** sets out the values obtained. The column on the left contains the variables corresponding to the construction elements that make up the combinations of the building envelope. The following four columns show the value of the impacts due to the retrofit process and the costs of substitution or reparation of the construction elements. Their corresponding units are expressed by square meters of built assembly.

Table 4: Impacts a result of the construction assembly in the energy retrofit phase

VARIABLES		ENERGY RETROFIT PHASE			
Construction assembly		Costs		Impacts	
		Investment in retrofit	Consumption cost	Emission	Primary energy consumption
		(€/m²)	(€/m²)	(kg CO_2/m²)	(kWh/m²)
C1		51.95	16.11	112.45	88.93
C2		47.94	18.00	46.34	99.36
C3		34.64	17.38	53.55	95.96
F1		11.14	2.79	4.63	15.39
F2		11.14	2.79	4.63	15.39
F3		22.91	2.22	23.14	12.24
F4		18.04	4.36	5.78	24.06
F5		30.67	10.68	12.77	58.98
H1		348.24	272.54	659.06	1504.71
H2	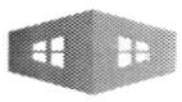	336.24	162.66	359.16	898.07
H3		415.38	55.02	71.88	303.75

The indicators are obtained by calculating a common factor for each group of construction assemblies and for each impact that corresponds to the lowest value of the impact in each group that is shaded in Table **4**. The factors or coefficients of each impact are then obtained by dividing their real value by the lowest impact value in each group, as is set out in Table **5**.

Table 5: Factors obtained for each impact in the energy retrofit phase

Construction assembly		ENERGY RETROFIT PHASE			
		Investment in retrofit Factor	**Consump cost Factor**	**Emission CO_2 Factor**	**Primary energy consump Factor**
		F_{cub}^{man} €/m²	F_{cub}^{man} €/kW	F_{cub}^{man} CO_2	F_{cub}^{man} kW·h
C1		1.50	1.00	2.43	1.00
C2		1.38	1.12	1.00	1.12
C3		1.00	1.08	1.16	1.08
		F_{fac}^{man} €/m²	F_{fac}^{man} €/kW	F_{fac}^{man} CO_2	F_{fac}^{man} kW·h
F1		1.00	1.26	1.00	1.26
F2		1.00	1.26	1.00	1.26
F3	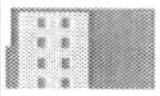	2.06	1.00	5.00	1.00
F4		1.62	1.97	1.25	1.97
F5		2.75	3.83	2.76	4.82
		F_{car}^{man} €/m²	F_{car}^{man} €/kW	F_{car}^{man} CO_2	F_{car}^{man} kW·h
H1		1.04	4.95	9.17	4.95
H2	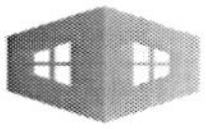	1.00	2.96	5.00	2.96
H3		1.24	1.00	0.20	1.00

The equations developed to calculate the indicators in the energy retrofit phase are then obtained. The value of the coefficients will be replaced in each case by the corresponding coefficient set out in Table **5**, according to the option chosen by the designer. The surface factor ($F_{s,cub}$; $F_{s,fac}$; $F_{s,car}$) is an item that the designer will have to calculate in each project, as well as the useful life (V_{Ures}; V_U).

Equivalent $C0_2$ emissions in the energy retrofit phase

$$Em_{eqCO_2}^{man} = \frac{V_{Ures}}{V_U}\{(46.34 \cdot F_{S,cub} \cdot F_{cub,CO_2}^{man}) + (4.63 \cdot F_{S,fac} \cdot F_{fac,CO_2}^{man}) + (71.88 \cdot F_{S,car} \cdot F_{car,CO_2}^{man})\} \quad (3)$$

Primary energy consumptions during the energy retrofit phase

$$Con^{man}_{ep\mathbf{kWh}} = \frac{V_{Ures}}{V_U}\{(88.93 \cdot F_{S,cub} \cdot F^{man}_{cubkWh}) + (12.24 \cdot F_{S,fac} \cdot F^{man}_{fackWh}) + (303.75 \cdot F_{S,car} \cdot F^{man}_{car\,kWh})\} \quad (4)$$

Investment costs in the retrofit phase

$$Cost^{man}_{€/m^2} = \frac{V_{Ures}}{V_U}\left\{(34.64 \cdot F_{S,cub} \cdot F^{man}_{cub\frac{€}{m}2}) + (11{,}14 \cdot F_{S,fac} \cdot F^{man}_{fac\frac{€}{m}2}) + (336{,}24 \cdot F_{S,car} \cdot F^{man}_{car\,€/m^2})\right\} \quad (5)$$

Costs of the energy consumed in the energy retrofit phase

$$Cost^{man}_{ep\,€/m^2} = \frac{V_{Ures}}{V_U}\{(16.11 \cdot F_{S,cub} \cdot F^{man}_{cubkWh}) + (2.22 \cdot F_{S,fac} \cdot F^{man}_{fackWh}) + (55.02 \cdot F_{S,car} \cdot F^{man}_{car\,kWh})\} \cdot cost_{kWh} \quad (6)$$

The equations for the use phase are obtained following a process similar to that which has already been developed for the first version model, as previously described, as the comfort conditions after the retrofit of existing buildings are similar to new-builds. The life span considered is the residual life span.

Creation of the Excel Matrix to Implement the Model

After the equations to calculate the indicators in the energy refurbishment phase were obtained, they were implemented in an Excel matrix. The equations for the use phase obtained in the first version model were also implemented in this matrix, as has been previously explained.

The foregoing allows different alternative designs to be chosen for the building envelope in different climate zones and orientations, so that the estimated values of the impacts generated in this phase during the rest of the life cycle of the building can be automatically displayed. There is a column to incorporate the Fs (Surface factor), being a value which corresponds to the proportional share of the surface area of each element in the envelope (roof, façade, carpentry) per square meter of usable floor area of the building.

The right-hand part of the matrix contains the columns of the results and is divided into two blocks, one for each of the phases of the life cycle of the building evaluated: the refurbishment phase subdivided into four columns and the use phase subdivided into six columns, corresponding to the impacts during the rest of the building life span once it has been retrofitted.

Each column contains an impact in its corresponding phase and shows the name of the indicator and the corresponding units according to the type of impact per square meter of usable floor area.

The designer can use this calculation matrix by selecting the type of dwelling in the first column and using the second column to indicate the surface factor, Fs. The construction assemblies available in this study are indicated by a letter and a number, which are identified with the corresponding construction solutions. Furthermore, the user can also select the climate zone as well as the building orientation.

The estimated value of the impacts is obtained because the equations developed for each indicator are linked to the corresponding cell of the calculation matrix (Fig. **3**). Columns belonging to the same indicator are identified with the same color in the different phases of the life cycle. The user obtains information regarding the value of the impacts in real time and can compare different design options. Thus, Fig. **3** shows the calculation matrix with the results obtained for the options that were selected.

CONCLUSION

A model has been developed to calculate the environmental impacts related to the materials used during the refurbishment phase, as well as the impacts linked to the envelope produced during the use phase of the building by reason of air conditioning installations. It also makes it possible to evaluate the economic cost of the impacts associated to its life cycle in terms of both investment and energy saving.

The calculation matrix allows the planner to know the value of the estimated impacts of each one of the possible envelope retrofit combinations by selecting different options on the matrix.

The obtained results have enabled us to determine by what proportion the impacts related to the retrofitting materials may be considered and estimated compared to the environmental impacts caused by the use of the installations during the use phase. Naturally, the higher the life span which is considered, the higher the impacts generated by the use phase. It is important to note that the CO_2 emissions due to the retrofitting materials represent approximately 16% of the emissions due to the use phase during a residual life span of 50 years, and that the energy consumption due to the retrofitting materials represents around 7% of the energy consumption during the 50 years use phase.

BUILD TYPE	VARIABLES: Surfase factor		VARIABLES: Constructive assemblies											VARIABLES: ORIENTATION		VARIABLES: CLIMATE ZONE	RETROFIT PHASE: Eq. CO_2 emissions $Em^{man}_{eqCO_2}$	RETROFIT PHASE: Energy consump Con^{man}_{epkWh}	RETROFIT PHASE: Energy costs $Cost^{man}_{ep€/m^2}$	RETROFIT PHASE: Retrofit costs $Cost^{man}_{€/m^2}$	USE PHASE: Eq. CO_2 heating emissions $Em^{uso}_{eq.CO_2cal}$	USE PHASE: Eq. CO_2 conditioning sions $Em^{uso}_{eq.CO_2ref}$	USE PHASE: Energy heating consump $Con^{uso}_{kWh\,cal}$	USE PHASE: Energy conditioning consump $Con^{uso}_{kWh\,ref}$	USE PHASE: Energy heating costs $Cost^{uso}_{ep€/m^2cal}$	USE PHASE: Energy conditioning costs $Cost^{uso}_{ep€/m^2ref}$
	Type	Fs	C1	C2	C3	F1	F2	F3	F4	F5	H1	H2	H3	NE	SE	B3	kg CO_2/m²	kWh/m²	€/m²	€/m²	kg CO_2/m²	kg CO_2/m²	kWh/m²	kWh/m²	€/m²	€/m²
SEMIDET		0,72		■										■		■	84,51	182,10	32,98	107,37	459,00	240,50	2.295,00	985,00	415,68	174,79
BLOCK		0,23				■								■		■										
DETACHED		1,00									■			■		■										
SEMIDET		0,72			■									■		■	65,48	152,03	27,54	88,49	320,50	215,00	1.609,50	933,50	291,52	169,08
BLOCK		0,23						■						■		■										
DETACHED		1,00										■		■		■										

Figure 3: Calculation matrix with the resulting values for each impact in the retrofit and use phases.

The impacts due to the retrofit process were obtained according to the proportionality of each element which constitutes the construction assembly which enables the designer to provide this information for the calculation and obtain multiple results depending on the specific design adopted.

Regarding the façades, it can be observed that the insulation thickness determines to a great extent the resulting impacts during the use phase. The best results are obtained for the façade assembly with thicker insulation, F3, both for heating and cooling. Comparing assemblies with the same materials combinations but with different insulation thickness (5 cm and 10 cm), it can be observed that whereas in the fabrication and installation phases the resulting values for the impacts remain similar, in the use phase of the assemblies with thicker insulation the CO_2 emissions fall by approximately 100 kg of CO_2/m^2 and energy consumption fall by around 250 $kW{\cdot}h/m^2$ with regard to the thinner insulation, considering a 50 year residual life span.

If the solutions are compared modifying only the type of roof, the combinations that belong to the inverted roof, C3, provide the best results due to the low maintenance required and good resistance to heat.

Regarding the type of carpentry, the impacts during the use phase do not vary significantly due to the frame materials described, but rather depend on the quality of the specific frame and glass selected for the envelope retrofit, defined by parameters such as transmittance, permeability, *etc*.

As final conclusion, it could be said that the developed model can facilitate the comparison and selection work of the designer, as a building sustainability manager, by means of the characterization of alternatives to carry out building envelope retrofits by combining different materials that form a part of a construction assembly, thus enabling the designer to deploy the impacts of each of the phases of the life cycle of the building and facilitating the easy detection of the most critical phase or the most critical impact.

ACKNOWLEDGEMENTS

This paper has been carried out within the context of the research project "Integrated design protocol for social housing retrofit and urban regeneration" (BIA2013-44001-R) within the Spanish Research, Development and Innovation Program "Society's Challenges", funded by the Spanish Ministry of Economy and Competitiveness.

CONFLICT OF INTEREST

The authors confirm that this chapter contents have no conflict of interest.

REFERENCES

[1] P. Huedo, "*La evaluación del impacto ambiental de la envolvente del edificio como herramienta de apoyo en fase de diseño aplicada a viviendas*", Ph. D. thesis, Universitat Jaume I, Castellón, July 16, 2014.

[2] International Energy Agency, "*Promoting Energy Efficiency investments*", 2008, [Online] Available; http://www.iea.org/publications/freepublications/publication/PromotingEE2008.pdf.

[3] M.J. Ruá, "*Métodos de valoración de viviendas desde la perspectiva medioambiental y análisis de costes*", Ph. D. thesis, Universidad Politécnica de Valencia, Valencia, July 1, 2011.

[4] A. Haapio and P. Viitaniemi, "A critical review of building environmental assessment tools", *Environmental Impact Assessment Review*, vol. 28, pp. 469-482, October 2008.

[5] P. Huedo and B. López-Mesa. Review of tools to assist in the selection of sustainable building assemblies. *Informes de la Construcción*, vol. 65, pp. 77-88, January-March 2013.

[6] M. Erlandsson and M. Borg, "Generic LCA-methodology applicable for buildings, constructions and operation services-today practice and development needs", *Building and Environment*, vol. 38, pp. 919-938, July 2003.

[7] T. Argüello and A. Cuchí, "Analysis of the environmental impact associated with the materials of construction used in the low cost houses of the program 10 x10 With Roof-Chiapas of the CYTED", *Informes de la Construcción*, vol. 60, pp. 25-34, January-March 2008.

[8] C. Alonso, I. Oteiza and J.García, "Criterios para la reducción de emisiones de gases de efecto invernadero en el proyecto de fachadas de edificios de Viviendas", In: *Proceedings of II Congreso Nacional Investigación en Edificación*, 2010.

[9] G. Gonzalo, S. Ledesma, V. Nota, C. Martínez, S. Cisterna, G. Quiñónez, G. Márquez, A. Tortonese and A. Garay, "Determinación y análisis de los requerimientos energéticos para el acondicionamiento térmico de un prototipo de vivienda ubicada en San Miguel de Tucumán", *Revista Avances en Energías Renovables y Medio Ambiente*, vol. 4, pp. 05.19-05.24, 2000.

[10] I. Oteiza and C. Alonso, "Análisis y revisión de herramientas para evaluación de la sostenibilidad de la construcción", In: *Proceedings of II Jornadas de Investigación en Construcción*, 2008, pp. 1149-1166.

[11] N. Mithraratne and B. Vale, "Life cycle analysis model for New Zealand houses", *Building and Environment*, vol. 39, pp. 483-492, April 2004.

[12] O. Ortiz, C. Bonnet, J. Bruno and F. Castells, "Sustainability based on LCM of residential dwellings: A case study in Catalonia, Spain", *Building and Environment*, vol. 44, pp. 584-594, March 2009.

[13] ITeC (Instituto de la Construcción de Cataluña), "*Programa para los proyectos y obras de construcción – TCQ*", Barcelona, [Online] Available; http://itec.es/programas/tcq/.

[14] BEDEC, "*Database of construction materials developed by ITeC*" (Instituto de la Construcción de Catalunya), Barcelona, 2013 version, [Online] Available; http://itec.es/nouBedec.e/.

[15] LIDER v1.0. "*Manual de usuario, Instituto para la Diversificación y Ahorro de la Energía (IDAE) y Dirección General de Arquitectura y Política de Vivienda*", 2007. [Online] Available; http://www.codigotecnico.org/web/galerias/archivos/ManualLIDER.pdf

[16] AICIA (Grupo de Termotecnia de la Escuela Superior de Ingenieros Industriales de la Universidad de Sevilla–), "*Calener-VYP. Viviendas y edificios terciarios pequeños y medianos. Manual de usuario*", Instituto para la Diversificación y Ahorro de la Energía (IDAE) y Ministerio de Vivienda, 2009. [Online] Available; http://www.minetur.gob.es/energia/desarrollo/EficienciaEnergetica/CertificacionEnergetica/DocumentosReconocidos/ProgramaCalener/CalenerVYP1/Manual_de_usuario.pdf.

[17] F. Nemry and A. Uihlein, "*Environmental Improvement Potentials of Residential Buildings (IMPRO-Building)*", Office for Official Publications of the European Communities, 2008. [E-book] Available;http://publications.jrc.ec.europa.eu/repository/bitstream/111111111/7070/1/reqno_jrc46667_jrc46667%5B1%5D%5B1%5D.pdf.pdf.

[18] S. Hernández Moreno, "Aplicación de la información de la vida útil en la planeación y diseño de proyectos de edificación", *Acta Universitaria*, vol. 21, pp. 37-42, May-August 2011.

[19] L. Ortega, "*Propuesta metodológica para estimar la vida útil de los sistemas constructivos de fachadas y cubiertas utilizados actualmente con más frecuencia en la edificación española a partir del método propuesto por la Norma ISO-15686*", Ph. D. thesis, Universitat Politécnica de Valencia, Valencia, December, 2012.

[20] Royal Decree 314/2006, of 17 March, which approved the Spanish Technical Building Code.

CHAPTER 5

A Structure for the Quantity Surveillance of Costs and Environmental Impact of Cleaning and Maintenance in Buildings

Alejandro Martínez-Rocamora*, Jaime Solís-Guzmán and Madelyn Marrero

Department of Building Constructions II, ETSIE, University of Seville. Av. Reina Mercedes, 4-a. 41012 – Seville, Spain

Abstract: The long duration of the crisis in the construction sector in Spain after the bursting of the 'real state bubble' caused an almost total cessation of building activity, except for building maintenance and rehabilitation works. Predictive, preventive and corrective maintenance are the basis for extending the service life of buildings before rehabilitation is needed, and once that point has been reached, decrease the required investment. In this study, a structure is proposed for the development of prices that collect the cost and environmental impact of cleaning and maintenance works, which arise following the guidelines of the Andalusia Construction Cost Database (ACCD). To this end, a new chapter that covers use and maintenance activities is generated. Being these periodic tasks, their periodicity plays an important role in quantifying cumulative costs. When carrying out the quantity surveillance, the unitary costs, which initially seemed insignificant, increase considerably due to the high number of repetitions. As a result of this work, foundations for the development of the mentioned new chapter of the ACCD have been settled, with a solid structure that fits the existing one. Costs are structured intuitively by elements and uses of rooms, allowing a certain ease in locating a particular price. Moreover, their codes are designed to indicate somehow the periodicity with which tasks are performed. Finally, the designed structure and some examples of prices are shown, which additionally include environmental impact data.

Keywords: Carbon footprint, cleaning, costs, coding system, conservation, corrective, environmental impact, finishes, hotels, life-cycle costs, maintenance, periodicity, preventive, predictive, quantity survey, renovation, repair, replacement, revision, service life, structure.

INTRODUCTION

The long duration of the crisis in the construction sector in Spain after the bursting of the 'real state bubble' in late 2007 or early 2008 caused an almost total

***Corresponding author Alejandro Martínez Rocamora:** Department of Building Constructions II, ETSIE, University of Seville. Av. Reina Mercedes, 4-a. 41012 – Seville, Spain; Tel +34 675041010; E-mail: rocamora@us.es

cessation of building activity, except for building maintenance and rehabilitation works. During the years prior to that bursting, in Spain, there was a proliferation of small businesses, both architectural studios and construction companies. After six years, many of these companies have disappeared and the 'survivors' are forced to seek and acquire new capabilities to stand out from the rest. Among these new possibilities of expertise, life cycle cost assessment and environmental impact quantification, very close to each other, are included.

Life cycle cost assessment, whose general methodology is specified in ISO 15686-5: 2008 [1], involves expanding horizons in quantifying the costs of a building. Instead of sticking to the construction phase, a building's life cycle begins with the design phase and ends with its demolition or restoration, and in this time span are included the costs of operation, maintenance, repair or replacement of elements, and finally cleaning [2]. Operational costs refer to the supply of electricity, water (drinking water and purification), and liquefied gases. Maintenance includes a large number of periodic work of revision and conservation of installations and constructive elements, and in case a damage is detected, their repair or replacement. These tasks are usually referred to as predictive, preventive and corrective maintenance, which are the basis for extending the service life of buildings before rehabilitation is needed, and once that point has been reached, decrease the required investment.

If there is one factor that stands out among these as far as workload and frequency is concerned, that is cleaning. Although it also has a big impact on residential and public buildings, cleaning acquires maximum importance in hotels, where it becomes an essential work on which depends the smooth running of the business. In this scope, the frequency with which each activity is performed grows exponentially, leading us to consider how much they cost and how much environmental impact is generated by these activities over the service life of the building.

There exist several tools for calculating life cycle costs, as BCIS Online, from the Royal Institute of Chartered Surveyors [3], which requires a license fee; Building Life Cycle Cost Program, from the National Institute of Standards and Technology [4, 5]; or ECONPACK, from the Corps of Engineers of the United States Navy [6], the latter being restricted to the general public. Among these, the only one that specifically says contemplate cleaning costs is BCIS Online, and obtaining a license is limited to businesses.

The shortage of available tools is partly due to the breakneck speed that the construction sector reached in the boom years, when the maintenance that constructed buildings may subsequently require cared little or nothing. In the research field, the study of cleaning in buildings is limited to the performance with different flooring types [7, 8], the estimation of cleaning costs for office buildings by regression of their total operating costs [9], analysis of maintenance in hospitals and dwellings [10, 11] and performance of operations in hotels [12], and finally various ergonomic and physical burden studies of professional staff in this sector [13-15].

In this study, the development of prices that collect costs and environmental impact of cleaning and maintenance tasks is proposed, which arise following the guidelines from the Andalusia Construction Cost Database (ACCD), with free access [16]. In the next section, some of the main features of the ACCD's structure, which is used as basis for the development of the structure and coding for a new chapter, are shown.

MATERIALS AND METHODOLOGY

The Andalusia Construction Cost Database (ACCD) has established a systematic classification consolidated over the years. It follows a structure in which (BCs) Basic and Auxiliary Costs (ACs) combined make up the Unitary Costs (UCs) or items that eventually make up the Production Total Cost (Fig. **1**) [17].

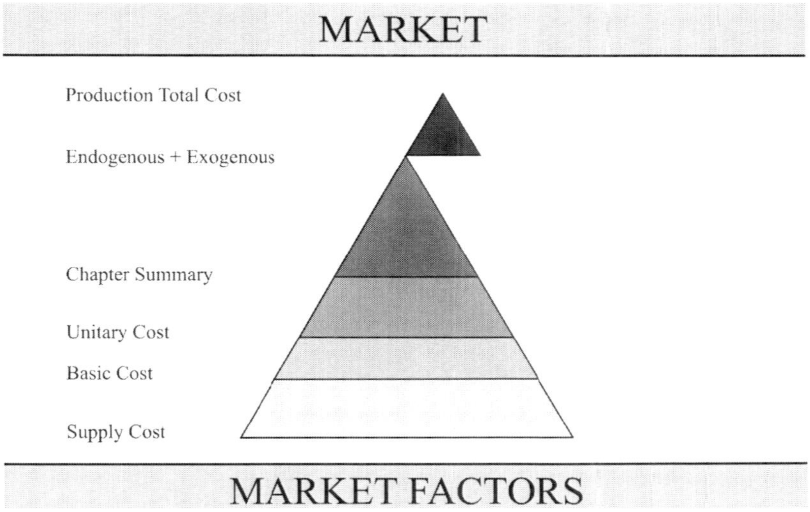

Figure 1: Pyramidal Costs Structure (Marrero and Ramírez de Arellano, 2010).

The coding of UCs is based on this classification system, with 10 alphanumeric characters: 2 figures representing the stage of the construction process, 3 letters to specify the category of the constructive element, and five more digits for uniquely numbering the item or entry. With this system a one-one correspondence between codes and entries is established [18].

In this coding system, a broken down Simple Unitary Cost (SUC) follows the structure in Table **1**, with its corresponding code, the unit in which it is expressed, a short description of the cost concept, its detailed full description, the measurement criterion to be used, and the list of costs that compose it, defining their quantity, unitary price and total cost, with which eventually direct, indirect, and overall costs are calculated.

Table 1: Structure of a simple unitary cost in the ACCD

<table>
<tr><th>SUC CODE</th><th>unit</th><th colspan="4">SHORT DESCRIPTION</th></tr>
<tr><td colspan="6">Full description.</td></tr>
<tr><td colspan="6">Measurement criterion.</td></tr>
<tr><td>CODE</td><td>UNIT</td><td>CONCEPT</td><td>QUANTITY</td><td>PRICE</td><td>COST</td></tr>
<tr><td>C1 CODE</td><td>unit</td><td>C1 SHORT DESCRIPTION</td><td></td><td></td><td></td></tr>
<tr><td>C2 CODE</td><td>unit</td><td>C2 SHORT DESCRIPTION</td><td></td><td></td><td></td></tr>
<tr><td></td><td></td><td></td><td colspan="2">Direct Costs</td><td></td></tr>
<tr><td></td><td></td><td></td><td>%</td><td>IC</td><td></td></tr>
<tr><td></td><td></td><td></td><td colspan="2">TOTAL</td><td></td></tr>
</table>

In the systematic classification of the ACCD, all the stages of the building process are covered in separate chapters, from demolitions and previous works to paintings and equipment, as well as chapters focused on urbanization, waste management and health and safety, the latter being number 19.

The methodology will consist on the development of a proposal for structuring a new chapter of the ACCD dedicated to the use and maintenance of buildings. For this, first, a new way of using the coding system to identify the different tasks is defined. Subsequently, a method to calculate the costs and environmental impact of a comprehensive set of tools or products used for these tasks is developed. Finally, examples of development of BCs, SUCs and Functional Unitary Costs (FUCs) are presented.

PROPOSED STRUCTURE

Since all existing chapters are related to the construction phase, it seems appropriate that the use and maintenance, being another phase of the building's life cycle, to be covered in a new chapter after the existent ones, in this case number 20. This chapter, as you can tell by its name, not only refers to cleaning tasks, but also maintenance, such as periodic revisions, repairs or replacements, and actions for the conservation of the proper functioning of constructive elements.

Assigning the following three characters is a first problem in the structure, since the use of the three of them to define the category of the constructive element, as it is done in previous chapters, causes the groups of entries remain quite small, being usually five (conservation, cleaning, revision, repair and replacement) per group. To avoid this situation, it is proposed that in this chapter, constructive elements should be defined by the first two letters, thus forming more general sets with a larger number of entries within (see Table **2**), while the third letter specifies the type of task to be done (Table **3**).

Table 2: Coding for subchapters and sections

Subcapítulos/*Subchapters*		**Apartados/*Sections***	
20A	AISLAMIENTOS */ISOLATIONS*	20AV	Aislamientos vistos/*Visible isolations*
		20AW	Varios/*Various*
20C	CIMENTACIONES */FOUNDATIONS*	20CL	Losas/*Concrete slabs*
		20CM	Muros pantalla/*Retaining walls*
		20CW	Varios/*Various*
20D	DIVISIONES INTERIORES */INTERIOR PARTITIONS*	20DB	Barandillas/*Railings*
		20DD	Divisiones/*Partitions*
		20DM	Mamparas/*Screens*
		20DP	Puertas/*Doors*
		20DW	Varios/*Various*
20E	ESTRUCTURAS */STRUCTURES*	20EA	Acero/*Steel*
		20EF	Forjados/*Floor slabs*
		20EH	Hormigón/*Concrete*
		20EM	Madera/*Wood*
		20EW	Varios/*Various*
20F	FACHADAS */FAÇADES*	20FC	Carpintería y protección/*Windows, doors and protection*
		20FP	Paredes/*External walls*
		20FW	Varios/*Various*
20I	INSTALACIONES */INSTALLATIONS*	20IA	Aire acondicionado/*Air conditioning*
		20IC	Calefacción/*Heating*
		20IE	Electricidad/*Electricity*

Table 2: contd….

		20IF	Fontanería/*Plumbing*
		20II	Iluminación/*Lighting*
		20IK	Comunicación/*Communications*
		20IL	Gases y licuados/*Gases and liquified fuels*
		20IM	Electro-mecánicas/*Electro-mechanical*
		20IN	Energía solar/*Solar energy*
		20IP	Protecciones/*Protections*
		20IR	Evacuación de residuos/*Waste disposal*
		20IS	Saneamiento/*Sewerage*
		20IV	Ventilación/*Ventilation*
		20IW	Varios/*Various*
20M	MOBILIARIO /*FURNITURE*	20MA	Aseos y baños/*Toilets and bathrooms*
		20MC	Cocinas/*Kitchens*
		20MD	Dormitorios/*Bedrooms*
		20ME	Estanterías y armarios/*Shelves and closets*
		20MK	Decoración/*Decoration*
		20ML	Lámparas/*Lamps*
		20MM	Mesas/*Tables*
		20MS	Sillas, sofás y sillones/*Chairs, sofas and armchairs*
		20MW	Varios/*Various*
20P	ESPACIOS INTERIORES /*INTERIOR SPACES*	20PA	Aseos y baños/*Toilets and bathrooms*
		20PB	Bibliotecas y aulas/*Libraries and classrooms*
		20PC	Cocinas/*Kitchens*
		20PD	Dormitorios/*Bedrooms*
		20PE	Escaleras y rampas/*Stairs and ramps*
		20PK	Comedores y cafeterías/*Dining rooms and cafeterias*
		20PL	Laboratorios/*Laboratories*
		20PM	Almacenes/*Store rooms*
		20PO	Oficinas y despachos/*Offices*
		20PP	Pasillos y distribuidores/*Hallways*
		20PS	Salones y salas/*Living, meeting and conference rooms*
		20PV	Vestíbulos/*Halls*
		20PW	Varios/*Various*
20Q	CUBIERTAS/*ROOFS*	20QH	Horizontales/*Horizontal*
		20QI	Inclinadas/*Sloped*
		20QW	Varios/*Various*

Table 2: contd….

20R	REVESTIMIENTOS /*COVERINGS*	20RA	Aplacados/*Platings*
		20RC	Continuos/*Continuous*
		20RL	Ligeros/*Lightweight*
		20RP	Pinturas/*Paintings*
		20RS	Suelos/*Floors*
		20RT	Techos/*Ceilings*
		20RW	Varios/*Various*
20U	URBANIZACIONES /*URBANIZATIONS*	20UA	Alcantarillado/*Sewage system*
		20UD	Instalaciones deportivas y de ocio/*Sports facilities*
		20UE	Electricidad/*Electricity*
		20UJ	Jardinería/*Gardening*
		20UM	Mobiliario/*Furniture*
		20UP	Pavimentos/*Pavements*
		20UR	Sistemas de riego/*Irrigation systems*
		20US	Suministro de agua/*Water supply*
		20UW	Varios/*Various*
20W	VARIOS/*VARIOUS*	20WW	Varios/*Various*

The different types of tasks are divided into conservation tasks, meant to be adjustments or cleaning works which allow the conservation of the item's proper functioning, this being also known as preventive maintenance; cleaning tasks, with which the cleanliness of the visible surfaces of the building is kept; periodic revisions of constructive elements, as part of the so-called predictive maintenance; replacements and repairs, or corrective maintenance; and finally the 'Various' group.

Table 3: Coding for groups (type of task)

20**C	Conservaciones/*Conservation*
20**L	Limpiezas/*Cleaning*
20**R	Revisiones/*Revisions*
20**S	Sustituciones y reparaciones/*Replacements and repairs*
20**W	Varios/*Various*

It is usual that one of the first things an architect wants to know before a rehabilitation is what kind of maintenance has been made to the building, also because the more maintenance a building has had during its service life, the less work will be needed for its rehabilitation. In fact, sometimes maintenance might be thought of as kind of a prolonged-in-time rehabilitation.

As for the last 5 digits of the code, in addition to uniquely numbering the various items, it is intended that at least the first digit indicates somewhat the frequency with which tasks are performed, thus making easier to find an entry. In the present case, focusing for the moment on cleaning tasks, the cleaning of building elements or furniture shall be calculated without a specific frequency, forming SUCs. By contrast, periodic cleaning of different types of rooms are defined as broken down FUCs, in which some of the above are included. The values chosen to represent the periodicities are presented in Table **4**.

The development of SUCs of cleaning is based on BCs corresponding to labour, materials and tools used. For labour, the cost of a special labourer is used, which represents a worker in a case in which there is not a cleaning crew with a person in charge, while for materials and tools there are two possible options. The first one is to specify, for each task, the exact products that will be used, which restricts the flexibility of entries but gives greater accuracy in the calculation of their cost and environmental impact. The second option involves calculating the cost for a whole cleaning trolley, including all products and utensils carried in it. This option simplifies the calculations and the development of costs, but carries inherent in the breakdown of the different products a factor of frequency of cleaning some kinds of elements, or amount of product used in each use. For example, if we consider that, within the cleaning trolley, a 750 ml bottle of glass cleaner is spent in a month and a 750 ml bottle of bath descaler must be renewed each three months, we are indirectly saying that glass is cleaned more often (windows, mirrors, *etc.*) than bathrooms, or otherwise, that much more glass cleaner than descaler is used per cleaned surface.

Table 4: Coding for the frequency of tasks

Code	**Frequency**
00000	< Daily
10000	Daily
20000	Weekly
30000	Monthly
32000	Bimonthly
33000	Quarterly
36000	Semiannual
40000	Annual
42000	Biennial
43000	Triennial

Table 4: contd….

44000	Every 4 years
45000	Quinquennial
50000	> Quinquennial
90000	Undefined

This work is part of a larger one in which a student accommodation building belonging to the University of Seville is used as a case study, which is considered to have some operating patterns similar to those of a hotel. Because of this, data on cleaning frequencies of items and rooms, as well as products and utensils used by staff and their wear or performance, have been obtained through the Cleaning Unit of the University of Seville.

Below, the results obtained for BCs are shown, as well as the development of an example of SUCs and a FUCs wherein the structure and coding explained here apply.

RESULTS AND DISCUSSION

As it was mentioned before, BCs of products and cleaning utensils can be calculated in two ways: individually, or as a whole forming a cleaning trolley. In either case, the auxiliary table designed to develop these prices includes every item usually contained in a cleaning trolley, with performance or wear for each one, and its total, annual, and hourly cost. Regarding environmental data, if the analysis is aimed at the Carbon Footprint (CF) indicator, we need to know the material of which each element is made, its emission factor of CO_2 equivalent, weight, and finally, its CF per year and hour (Table **5**). All environmental data come from reports for the Ecoinvent database v2.1 for Life Cycle Assessment of products and processes [19].

If the cleaning trolley is treated as a whole in order to calculate the various costs, in Table **5**, a total hourly cost of 0.57 €/h, and a total hourly emissions of 0,135 kg of CO_2 equivalent per hour are obtained. This can be done in a similar way with the tools of a general maintenance technician, a carpenter, a refrigeration technician or an electrician.

It is to be clarified that CO_2 equivalent is the unit used by the CF indicator, consisting of a unification of the various greenhouse gases emitted into the atmosphere (CH_4, NO_x, CFC, O_3, *etc.*), through conversion coefficients based on the global warming potential that each one has compared to CO_2 [20].

Table 5: Cost and environmental impact of cleaning products and utensils

CLEANING TROLLEY									
ITEM	**RENOV. (yr)**	**COST (€)**	**ANNUAL COST (€/yr)**	**HOURLY COST (€/h)**	**MAIN MATERIAL**	**EMISSION FACTOR (kg CO_2 eq/kg)**	**WEIGHT (kg)**	**ANNUAL EMIS. (kg CO_2 eq/yr)**	**HOURLY EMISSIONS (kg CO_2 eq/h)**
Cleaning trolley	5.000	137.09	27.42	0.02	Polypropylene	1.97	14.000	5.516	0.004
Wide mop	0.043	4.83	112.33	0.07	Cotton	1.86	0.515	22.277	0.015
Dust clothes (3)	0.167	3.70	22.17	0.01	Polyester	7.54	0.117	5.283	0.003
Buckets (2)	1.000	15.92	15.92	0.01	Polypropylene	1.97	0.350	0.690	0.000
Mop or broom sticks (4)	0.500	14.52	29.04	0.02	Aluminium	8.55	0.984	16.826	0.011
Mops (2)	0.083	2.42	29.16	0.02	Cotton	1.86	0.440	9.860	0.006
Broom	0.250	1.20	4.80	0.00	PET	2.75	0.225	2.475	0.002
Dustpan	1.000	1.44	1.44	0.00	Polypropylene	1.97	0.135	0.266	0.000
Big garbage bag	0.004	0.16	40.75	0.03	Polyethylene (LDPE)	2.10	0.010	5.250	0.003
Trash bags	0.009	0.91	101.11	0.07	Polyethylene (LDPE)	2.10	0.150	35.000	0.023
Wiper blade	2.000	10.32	5.16	0.00	Chromium steel	5.23	0.211	0.552	0.000
Wiper blade rubber	0.250	1.72	6.88	0.00	Synthetic rubber	2.63	0.070	0.736	0.000
Wiper tool	2.000	11.35	5.68	0.00	Polypropylene	1.97	0.120	0.118	0.000
Wiper tool fleece	1.000	6.46	6.46	0.00	Cotton	1.86	0.050	0.093	0.000
Disposable gloves	0.004	0.01	1.63	0.00	Polyethylene (LDPE)	2.10	0.005	2.625	0.002
Paper hand towel roll 120m	0.065	2.71	41.69	0.03	Paper	1.35	1.512	31.403	0.021
Ostrich feather duster	0.250	5.45	21.80	0.01	Polypropylene	1.97	0.055	0.433	0.000
Antistatic duster	0.250	1.33	5.32	0.00	Polypropylene	1.97	0.115	0.906	0.001
Bleach (5L)	0.043	6.62	153.95	0.10	Surfactants (15%)	2.18	0.799	40.495	0.027
Neutral cleaner (1.5L)	0.043	2.24	51.98	0.03	Surfactants (15%)	2.18	0.226	11.441	0.008
Bath descaler (750ml)	0.250	2.77	11.09	0.01	Surfactants (15%)	2.18	0.113	0.981	0.001
Floor cleaner (1L)	0.022	1.50	68.05	0.04	Surfactants (15%)	2.18	0.225	22.295	0.015
Furniture cleaner spray (450ml)	0.167	4.68	28.02	0.02	Surfactants (15%)	2.18	0.068	0.881	0.001
Mop cleaner spray (750ml)	0.250	2.15	8.60	0.01	Surfactants (15%)	2.18	0.080	0.697	0.000
Soapy cleaner (1L)	0.167	3.52	21.05	0.01	Surfactants (15%)	2.18	0.231	3.011	0.002
Glass cleaner (750ml)	0.083	0.90	10.81	0.01	Surfactants (15%)	2.18	0.113	2.955	0.002
Ink remover (750ml)	0.043	3.65	84.88	0.06	Surfactants (15%)	2.18	0.114	5.761	0.004
Bath gel filler (5L)	0.065	5.12	78.77	0.05	Surfactants (15%)	2.18	0.825	27.669	0.018
Air freshener (1L)	0.083	4.36	52.48	0.03	Surfactants (15%)	2.18	0.225	5.910	0.004

With these data and following the structure of the ACCD, the BC for the cleaning trolley is framed in Family H: Herramientas (Tools), subfamily L: Limpieza (Cleaning) (Table **6**), followed by a 5-digit numbering that would allow the inclusion of individual costs for all the products contained in the cleaning trolley.

As a final note, although it has not been stated in the following tables, costs are expressed in Euros, while CF is measured in kilograms of CO_2 equivalent.

Table 6: Basic Cost for a cleaning trolley

HL00100	**h**	**CLEANING TROLLEY**	
Quantified the hours used.			
		Direct Costs	0.57
		Carbon Footprint	0.135

SUCs have certain peculiarities in terms of data to be specified because of their direct influence on performance. In the case of floor, for example, the type of coating, with the presence of joints or not, and the percentage of area that is occupied by furniture elements are determinant to perform cleaning on more or less time. The same happens with wall finishes, where the difficulty lies in the coating roughness (smooth, embossed or textured walls, very rugged, *etc.*), or with windows, with more or less difficulty for cleaning the outer face or for the removal of the leaves. Thus, in Table **7** an example of SUC for sweeping of tiled floor with obstacles occupying between 10 and 30% of the total surface, which is common in public bathrooms, is shown.

Table 7: Simple Unitary Cost for sweeping of tiled floor, with its surface being occupied by obstacles between 10 and 30%

20RSL90012	**m^2**	**SWEEPING OF TILED FLOOR SURF. OC. 10%<S<30%**					
Sweeping of tiled floor. with its surface occupied by obstacles between 10 and 30%.							
Quantified the cleaned surface.							
CODE	**UNIT**	**CONCEPT**	**QUANTITY**	**PRICE**	**COST**	**EMISSIONS**	**CF**
HL00100	h	CLEANING TROLLEY	0.004	0.57	0.00	0.135	0.001
TP00100	h	SPECIAL LABOURER	0.004	18.28	0.07	0.518	0.002
			Direct Costs		0.07		0.003
			11%	**IC**	0.01		
			TOTAL		**0.08**		**0.003**

As it can be seen, to sweep a square meter of tiled floor, a time of about fourteen seconds has been estimated, which affects the usage time of the cleaning trolley and the work of the Special Labourer, in this case cleaning staff. These low figures make the final cost to be only 8 cents per square meter with 3 grams of CO_2 equivalent emitted, but we shall see what these values that a priori seem insignificant become when we calculate the number times that this task is performed.

In Table **7**, it also can be noticed that the assigned code indicates that the activity belongs to chapter 20, Use and Maintenance, subchapter R: Revestimientos (Coatings), section S: Suelo (Floor), Group L: Limpieza (Cleaning). The first two letters are used to classify the item in question, while the third one places it in the group of cleaning works. Meanwhile, the first of the last 5 digits of the code belongs to the category of tasks with an undefined frequency.

At a third level, and focusing on the building under study, we start with the development of FUCs. Functional Costs represent tasks that are carried out periodically, and are associated to rooms with a particular use. For example, once every day, the floor at the hotel corridors will be swept, and dust from the tables that may be in the corridor will be cleaned. Notice that, in wanting to synthesize different tasks by type of room, the area factor is lost. However, it is possible to define ranges of floor space, or in some cases capacity, where yields do not change drastically. Thus, different Functional Costs are created for assembly halls with capacity for less than 50 people, 50 to 100, and more than 100, as well as offices with less than 20 m^2 or between 20 and 40 m^2. In each of these cases both floor and wall surface, or the furniture elements contained in the room vary.

Table **8** presents a FUC designed to collect those cleaning tasks running 3 times a day in public bathrooms with a certain size, which fit a number of sanitary appliances, trash bins, and other items. It can be observed that the quantities used for the measurement of SUCs that compose it are half of the values that determine the defined intervals. For example, calculations are made considering 22.5 m^2 of floor (average between 15 and 30 m^2), and 8 sanitary appliances (6 to 10), which is the number of appliances that are estimated to fit in a public bathroom that size. Furthermore, the first SUC included in the breakdown is the one that was shown in Table 7. Regarding frequency, in other publications focused on cleaning floors of residential buildings, a weekly basis is assigned whether for the vacuuming or sweeping and mopping for all flooring types [8], or three different frequencies are set for each work generating three scenarios on cleaning habits of the occupants [7]. In our case, much higher frequencies are used because of the level of demand that a hotel business requires.

In relation to costs, it is important to note that in the column labeled 'PRICE', the unitary amount of the corresponding SUC is introduced regardless indirect costs (IC), as they will be computed in the FUC and otherwise there would be a double counting. The final result for the activity with code 20PAL04001 is 9.39 € and 0.301 kilograms of CO_2 equivalent, to be accounted for three times a day. If instead of a tiled floor with joints whose surface is occupied between 10 and 30%, there were a totally plain floor occupied by more than 70% surface, as would be the case in an assembly hall, the economic cost of cleaning with the same frequency would increase by 1.25 €, and emissions by 0.042 kg CO_2 eq, which helps to get an idea of the influence each SUC has on the Functional Cost. This new amount makes a big difference in the result for the SUC we saw in Table **7**, but its real importance will show up when calculating its total cost during the whole life cycle of the building.

Suppose that it is a hotel, which is running 365 days a year, where logically, cleanliness is one of the cornerstones of the business proper functioning. Customers have hygiene requirements, and not performing these tasks strictly may involve business decline. Running cleaning of a public bathroom located somewhere in the hotel costs, in this case, 9.39 €, which multiplied by 3 times a day and 365 days a year, become 10,282.05 € a year. It is quite significant now. Similarly, annual emissions amount to 329.595 kg CO_2 eq.

Table 8: Example of Broken down Functional Unitary Cost for cleaning 3 times a day in public bathroom

20PAL04001	**u**	**CLEANING 3 TIMES A DAY IN PUBLIC BATHROOM 15M²<S<30M²**					
Cleaning three times a day in public bathroom with a useful surface between 15 and 30 m2. including floor sweeping and mopping. trash emptying. and cleaning of 6 to 10 sanitary appliances.							
Quantified the cleaned unit three times a day.							
CODE	**UNIT**	**CONCEPT**	**QTY.**	**PRICE**	**COST**	**EMIS.**	**CF**
20RSL90012	m²	SWEEPING OF TILED FLOOR SURF. OC. 10%<S<30%	22.500	0.07	1.58	0.003	0.057
20RSL90062	m²	MOPPING OF TILED FLOOR SURF. OC. 10%<S<30%	22.500	0.08	1.80	0.003	0.065
20MWL90001	u	TRASH EMPTYING	1.000	0.15	0.15	0.005	0.005
20MAL90001	u	CLEANING OF SANITARY APPLIANCE	8.000	0.62	4.96	0.022	0.174
			Direct Costs		8.49		0.301
			11%	**IC**	0.90		
			TOTAL		**9.39**		**0.301**

To calculate the total cumulative costs over the service life of the building it would be necessary to estimate inflation (or discount rate) in order to update annual costs. In the same way, cumulative emissions would have a variation due to changes in the power generation methods, so patterns or trends should be established for the CF per unit of energy consumed.

With these magnitudes of results, we can imagine the importance that control and management of costs and environmental impact may have for the hotel management team. Small details like the choice of the flooring material, the type of windows installed, or the instruments used by the cleaning staff, can make significant differences in the costs and profitability of the business, or even to promote a more sustainable building.

CONCLUSION

In this paper, a robust and structured methodology for the generation of a new chapter of the Andalusia Construction Cost Database that includes the use and maintenance of buildings is set. Specifically, a suitable method has been described for the development of cleaning costs, with descriptive codes for the various categories in which they are included, as well as a method for encoding or classify the periodicities with which tasks are performed.

The developed costs show expected results, with small amounts in Simple Unitary Costs, which increase at the level of Functional Costs, but that become significant quantities when the number of repetitions and the service life of the building is applied. In addition, two columns have been added to the existing structure to collect emissions per unit and total Carbon Footprint for each task.

These estimates, which managers of hotels certainly might already be doing internally more or less accurately, had not been addressed to date from the point of view of a project's bill of quantities. The methodology described and the work here announced will certainly constitute a leap in Life Cycle Cost assessment.

Moreover, the knowledge of this information from the design phase of the building is considered useful, being able to weight up different possibilities for coatings or the layout of rooms, and check the variations that may occur in their costs and environmental impacts. Similarly, in the case of private residential buildings, their promoters or end users will be interested in obtaining extra information that will allow them to improve decision-making on coatings, finishes or materials, which will affect them during the complete service life of their homes. On the same line, being able to calculate the economic and environmental profitability of a building by using the costs included in the designed new chapter of the ACCD seems achievable and attractive. It also might be possible and interesting to evaluate if the cumulative costs of a regular basis maintenance would be smaller than the costs of rehabilitation if no maintenance at all had been made to the building.

ACKNOWLEDGEMENTS

Spanish Ministry of Innovation and Science, for funding the R&D project: Evaluation of the ecological footprint of construction in the residential sector in Spain (EVALHED: BIA2011-25812). 2012-2014.

Public Works Agency of Andalusia Government, for funding the project: Ecological Footprint of building rehabilitation: economic and environmental viability (HEREVEA: GGI3003IDIS). 2015.

CONFLICT OF INTEREST

The authors confirm that this chapter contents have no conflict of interest.

REFERENCES

[1] "*Buildings and constructed assets -- Service life planning -- Part 5: Life-cycle costing*", ISO 15686-5:2008.

[2] S. Fuller, "*Life-Cycle Cost Analysis (LCCA)*", NIST, National Institute of Standards and Technology, 2010.

[3] RICS, Royal Institution of Chartered Surveyors, "*BCIS Building Running Costs Online*", 2014. [Online] Available: http://www.rics.org/es/knowledge/bcis/online-products/bcis-building-running-costs-online/[Accessed Nov. 27, 2014].

[4] NIST, National Institute of Standards and Technology, "*Building Life Cycle Cost Program (BLCC)*", 2013. [Online] Available: http://www.energy.gov/eere/femp/building-life-cycle-cost-programs [Accessed Nov. 27, 2014].

[5] M. Davis, R. Coony, S. Gould, and A. Daly, "*Guidelines for Life Cycle Cost Analysis*". Stanford University Land and Buildings, 2005. [Online] Available; http://lbre.stanford.edu/sites/all/lbre-shared/files/docs_public/LCCA121405.pdf [Accessed Nov. 27, 2014].

[6] USACE, U.S. Army Corps of Engineers, "*Economic Analysis Package, ECONPACK*", 2014. [Online] Available: http://www.wbdg.org/tools/econpack.php [Accessed Nov. 27, 2014].

[7] E. Minne and J.C. Crittenden, "Impact of maintenance on life cycle impact and cost assessment for residential flooring options", *Int J Life Cycle Ass*, vol. 20, pp. 36-45, 2015.

[8] K. Allacker, "Environmental and economic optimization of the floor on grade in residential buildings", *Int J Life Cycle Ass*, vol. 17, pp. 813-27, 2012.

[9] C.A. Stoy and R. Johrendt, "Cleaning management of owner-operated real estate", *Facilities*, vol. 26, pp. 131-43, 2008.

[10] H. Gudat, "Facility management in German hospitals", *Health estate*, vol. 54, pp. 6-8, 2000.

[11] J.H.K. Lai, "Comparative evaluation of facility management services for housing estates", *Habitat Int*, vol. 35, pp. 391-7, 2011.

[12] C. Parkan, "Measuring the performance of hotel operations", *Socio-Econ Plan Sci*, vol. 30, pp. 257-92, 1996.

[13] H. Kalkis, Z. Roja, and V. Kalkis, "Physical load analysis in hotel cleaning work", *Agron Res*, vol. 12, pp. 843-50, 2014

[14] A.M. Seifert and K. Messing, "Cleaning up after globalization: An ergonomic analysis of work activity of hotel cleaners", *Antipode*, vol. 38, pp. 557-78, 2006.

[15] T. Scherzer, R. Rugulies, and N. Krause, "Work-related pain and injury and barriers to workers' compensation among Las Vegas hotel room cleaners", *Am J Public Health*, vol. 95, pp. 483-8, 2005.

[16] Junta de Andalucía, Base de Costes de la Construcción de Andalucía (BCCA). Banco de precios/Andalusia Construction Cost Database (ACCD). Costs database, 2014. [Online] Available: http://www.juntadeandalucia.es/fomentoyvivienda/portal-web/web/areas/vivienda/texto/7a0899c8-0038-11e4-8cc4-27ee69a25823 [Accessed Dec. 2, 2014].

[17] M. Marrero and A. Ramirez-de-Arellano, "The building cost system in Andalusia: application to construction and demolition waste management", *Constr Manage Econ*, vol. 28, pp. 495-507, 2010.

[18] J.L. Barón Cano, J. Conde Oliva, M. Osuna Rodríguez, A. Ramírez de Arellano Agudo, and J.A. Solís Burgos, Clasificación sistemática de Precios Básicos, Auxiliares y Unitarios/Systematic classification for Basic, Auxiliary and Unitary Costs. 2014. [Online] Available: https://ws147.juntadeandalucia.es/obraspublicasyvivienda/davwebviv/Banco%20de%20precios/BCCA_2014/BCCA_2014_junio/Clasi_Sistematica_2014.pdf [Accessed Dec. 2, 2014].

[19] Swiss Centre for Life Cycle Inventories, The ecoinvent Database, 2014. [Online] Available: http://www.ecoinvent.org/database/[Accessed Dec. 3, 2014].

[20] B.P. Weidema, M. Thrane, P. Christensen, J. Schmidt, and S. Løkke, "Carbon footprint", *J Ind Ecol*, vol. 12, pp. 3-6, 2008.

CHAPTER 6

Management of Living Urban Environments From an Active Ageing Approach: Advances in the Team Research (RE)Programa

Ángela Barrios[1,*], José C. Mariñas[1], Marta Molina[2], Francisco J. Lizana[1], Antonio Serrano[1] and Alberto Vilches[1]

[1]University of Seville, Department of Building Construction I, Reina Mercedes 2 Avenue, 41012, Seville, Spain and [2]University of Seville, Department of Building Structures and Geotechnical Engineering, Reina Mercedes 2 Avenue, 41012, Seville, Spain

Abstract: We are facing on the twenty-first century Andalusian cities significant changes in their social and economic structure, which proves the need of researching on policies that assure the right to housing. In the near future, we will find a difficult socioeconomic situation: the population of our cities and neighbourhoods will be composed of a majority of elderly people that will mostly live in obsolete dwellings that are not adapted to their housing requirements, considering moreover the aggravating fact that they will not be able to afford a refurbishment, upgrading or adaptation of the buildings. The public administration will not have enough resources to subsidise those works, neither will it be able to increase the offer of specialised housing to meet the expected demand in the medium and long term. In order to intervene in neighbourhoods through viable, efficient and adjusted to social, political and economic reality operations, the research team (Re)Programa (TEP-954), suggest the need for an integrated, multidisciplinary and multilevel analysis that considers four basic pillars, which are conservation of buildings, socio-economic situation, policies for energy saving and financing and management. We expose and analyse the objectives, criteria, analysis methodology, the case studies and the assessment tools developed in the project "(Re) habitation+ (Re) generation+ (Re) programming. Renovation and sustainable management of the Andalusian building stock. Management of living urban environments from an active ageing, gender and urban liveability approach", funded by the Department of Public Works of the Government of Andalusia, between 2013 and 2015.

Keywords: Ageing, building renovation, built environment, CO_2 savings, collective housing, economic levels, energy efficiency, financing tools, gerontology, guidelines, habitability, life satisfaction, life-quality, management, neighbourhoods, protocol, questionnaires, refurbishment, south Europe, sustainable development.

***Corresponding author Ángela Barrios:** Department of Building Construction I, School of Architecture, University of Seville, Seville, Spain; Tel: 0034 669 22 55 47; Fax: 954 55 70 18; E-mail: abarrios@us.es

Pilar Mercader-Moyano (Ed.)

INTRODUCTION

Demographic studies predict a significant increase in the population over 65 years of age in the near future, predicting it to double in 2040, and a decrease in the working-age population, which contributes to the financing of social security and the pension system, being necessary to design mechanisms and management or funding strategies to endow the citizen with some autonomy to make some adequacy and retrofitting operations viable in their housing and urban environments, in order to ensure optimal levels of quality of life.

The conservation of the city becomes complex with the ageing of the population. The funding difficulties for repairing their homes, and the physical ones to undertake a simple maintenance, lead to the progressive deterioration of the buildings and their surroundings. On the other hand, when elderly people begin to have limited mobility and autonomy, they tend to leave their homes and move to protected environments, generating in the city groups of empty properties in progressive deterioration.

People over 65 years of age form a special social group, because they are part of the support network working-age people rely on, because they generate wealth and jobs, as they are consumers of leisure activities, but also for their vulnerability: They are retirees, mostly women, who make use of the city and its infrastructure during the day in their dual role as users of educational, cultural, sport, and entertainment activities, and as caregivers of young children and people with reduced mobility.

The singularity of this population group justifies the need to design strategies that guarantee the implementation of social policies that are currently focused on two priority lines that promote:

- The active ageing, encouraging the participation of elderly people in professional, cultural, educational, and physical activities, ensuring a higher quality of life and reducing the negative effects of symptoms or disorders inherent to the ageing process. Elderly people who are active, physically and mentally, prevent the possible appearance of age-associated diseases.

- Ageing at home under conditions of dignity, valuing the person′s autonomy and providing support and state healthcare resources in case of need with programmes of home assistance for elderly people and their caregivers.

The Action Plans for energy saving and efficiency [1] are included in the political objectives of the European countries, including the 2020 target of saving 20% of the energy consumption, and establish the need to implement strategies for improving energy efficiency in the building sector, considering measures related to energy rehabilitation in existing buildings.

The law 8/2013 of June 26th for Rehabilitation, Regeneration and Urban Renovation [2], considers the obligation of the owners of residential flats, to have the Building Assessment Report available, in which the conservation status of the building and its adaptation to the basic conditions of universal accessibility are exposed, and the certificate of building energy efficiency is included.

The project we are presenting, entitled "(Re) Habitation + (Re) Generation + (Re) Programming. Recycling and sustainable management of the Andalusian building stock. Management of habitable environments from an active ageing, gender and urban liveability approach" of the research team "(Re)Programa", funded by the Public Works Agency of the Andalusian Government, proposes the development of intervention strategies with a strong integrated nature of social, environmental and economic measures that enable the implementation of policies to improve the energy efficiency of homes with the aim of improving comfort levels and ensuring a responsible energy consumption enabling:

- The implementation of comprehensive rehabilitation strategies in obsolete urban areas with a precarious level of income and facing a scenario where the reduction of public investment in grants joins the lack of private credit.

- The design of new housing ways to be included in the processes of urban recycling, to suit a scenario of an ageing population, achieving the reconciliation of the family life, caring for the elderly people and the working life.

- The revitalization of the consolidated city, solving housing situations of neglect and marginalization of the public space, reprogramming neighbourhoods and building containers and promoting the access to housing for the most vulnerable people.

The research team (Re)Programa (TEP-954) is composed of researchers from the University of Seville, of the Higher Technical School of Architecture, researchers from the University Pablo de Olavide, in the area of psychology and social work, and the School of Architecture at the University of Montevideo, in Uruguay, with a wide range of professionals in the field of architecture, construction, engineering and law, professionals and renowned companies with extensive experience in the field of energy rehabilitation, territorial and urban planning and liveability for elderly people, and the participation of scholarship holders from the University of Seville.

OBJECTIVES

The research we present has the main objective of generating tools to manage interventions at urban level, in neighbourhoods and buildings located in Andalusian cities, which will allow the most vulnerable collectives that live there to pay the costs needed to carry out a refurbishment and adaptation in order to enjoy the dwellings and neighbourhoods they know for the rest of their lives. This will be done considering diverse forms of management, finance and new cohabiting strategies that will allow these collectives to share the primary attention of elderly people and children at the same time as housework, transport, *etc*.

The objective of this paper is to show the integrated analysis methodology we design in this research project, reeling criteria and approaches from the fields of architecture and social sciences.

METHODOLOGY

We design a methodology based on four scenarios to perform an integrated, multilevel and multidisciplinary analysis:

- Conservation of buildings and urban regeneration.
- Socio-economic situation of the population. Needs and preferences of the elderly people.
- Policies for energy saving and efficiency of buildings and active ageing.
- Financing and management interventions of recycling and regeneration of the building stock.

We consider the following fields: city - neighbourhood - residential area - building - portal - housing, with a team of architects, building engineers and psychologists and from the basic objectives of:

1. Analyzing in deep the socioeconomic situation of the neighbourhoods, and gathering information about the needs, preferences and wishes of the elderly people that live in them.

2. Designing intervention protocols in neighbourhoods that integrate in a first phase the urban, architectonic and building detail analysis, as well as the establishment of the minimum requirements of habitability, accessibility and security, and secondly, designing intervention projects which define the necessary measures to be implemented and their costs.

3. Establishing financing methods for neighbourhood regeneration works and building refurbishment, and strategies to manage rents, maintain buildings and provide assistance to elderly people.

Phases and Tasks

We have developed the following phases and tasks:

A. ***Integrated Analysis.*** Characterization of buildings, urban environments, population, and urban regulations. This analysis considers the three following tasks:

A.1. Search and analysis of urban, architectural and socio-economic documentation.

A.2. Data collection on-site in locations selected.

Studies involving population:

a) Drafting and carrying out of surveys.

b) Generation of tables and graphics data.

c) Development of documents.

Analysis of urban environments and buildings:

a) Architectural, urban and building technology studies on-site.

b) Development of drawings, photographs and documents.

c) Drafting of brochures.

A.3. Information processing. It is developed through teamwork, obtaining an interdisciplinary point of view. A critical study of the results with experts on different areas is carried out.

B. *Projects.* We consider these three different tasks:

B.1. Diagnosis. From the data collected in the first phase, and once the study cases simulation models were generated and evaluated, the deficiencies and aspects that do not conform to legal requirements in relation to safety, habitability and energy efficiency are revealed.

B.2. Lines of Research and Intervention. The intervention proposals drawn by the research team were transferred to the collaborating companies, adjusting parameters and criteria in workshops. Work has been done together in this phase with the company "Territorio y Ciudad", and MRPR Architects, as well as with Txatxo Sabater, Margarita de Luxán and the Llornet-Llop-Pastor architectural office.

B.3. Quantification. Based on the proposal of three levels of intervention: minimum, medium and high, related to cost and energy savings percentages, the action lines are evaluated. MRPR Architects has the task of producing architectural proposals in preliminary projects at selected locations, making suggestions according to intervention budgets of funding management formulas in collaboration with the company "Territorio y Ciudad".

C. *Dissemination of Results*. In order to communicate and share the progress and results of the research project, we consider two tasks:

C.1. Generation of tools for disseminating results online: website [3], social networks (Facebook, Twitter), to show the advances of the project as well as the latest news about the team and current topics related to the research.

C.2. Development of conclusions and layout of the final report to be completed by the end of June 2015 with the publication of a book.

CASE STUDIES

The research key players are the architectonic and urban areas, and the neighbourhood inhabitants, thus the criteria considered for the selection of the building location has been the following:

- Regarding the inhabitants: the average socioeconomic level, taking into account pensioners and active professionals, with an average age over 65 years old, and with a significant percentage of people over 45 years old, whose dwellings are in property, or in public or private renting.

- Regarding the buildings: the locations with a certain degree of architectonic interest, with distinctive typologies of their construction period, within a range of age from 70 to 25 years and a suitable state of preservation, without serious structural problems were considered.

- Regarding the urban areas: the singularity and the intensity of use of the citizens, as well as the potential exploitation of those areas related to the cession to the works executor stakeholder, the green zones and community services, *etc*., were considered.

According to this criteria, the buildings studied by (Re)Programa are 11 and they are located in the cities of Seville (SE), Cordoba (CO), and Jerez de la Frontera (JE), as shown in Table **1**.

Table 1: List of building locations of the project (Re)Programa

Nº	Reference	Construction Year	Nº of Dwellings	Ownership
L01	SE- Los Remedios “Manzanas cuadradas”	1938-1943	324	Private
L02	SE- Tiro de Línea, “San Felipe”	1955	312	AVRA*
L03	SE- Los Remedios “Bloques Espiau”	1955	98 (686)	Private
L04	SE- Los Remedios “Bloques lineales”	1960	600	Private
L05	CO- San Martín de Porres	1960	1600	AVRA
L06	JE- S. Tomás de Aquino	1960	540	Private

L07	JE- Plaza Cabeceo	1968	332	Private
L08	SE- Triana, "Virgen de la Esperanza"	1969	513	Private
L09	SE- Santa Justa, "Las Huertas"	1980	640	AVRA
L10	SE- Lora del Río	1988	56	AVRA
L11	SE-San José de la Rinconada	1990	88	AVRA

*Note: AVRA, Agency of Housing and refurbishment of Andalusia, Regional Government of Andalusia.

INTEGRATED ANALYSIS INSTRUMENTS OF (RE)PROGRAMA

Analysis tools result of the research project from (Re)programa team have a great interest and are critical to academic and professional level. We design them in an integrated manner, in habitability workshops constituted to provide integrated from the beginning from the disciplines of architecture and civil engineering and social sciences. These tools are the Integrated Technical Inspection Sheet and Intervention Protocol, which are defined below:

- Integrated Technical Inspection Sheet. Tool for the characterization of the case study, designed to the scale of the city, neighbourhood, urban environment, building and housing. It considers administrative, architectural and technicians aspects [4], liveability surveys (accessibility and friendly use of spaces and mechanisms), and preferences in relation to the execution of works or cohabitation as formulas of coexistence issues. We describe the buildings studied, their structural typology, envelope, building systems, and their appropriateness to the requirements of Spanish standard (Technical Building Code [5]) in the Basic Documents: structural safety, security in case of fire, safety in use and accessibility, health, hygiene and protection of the environment, noise protection and energy saving. We employ a tool for assessing quality of life and vital and residential satisfaction validated indicators, from evaluation scales [6] of indicators of habitability that favour active ageing, and consider psychosocial aspects of residents, such as life satisfaction scale SWLS (Diener, 1985), Katz index of daily living activities (Katz *et al.*, 1963), the Lawton and Brody scale of instrumental activities of daily living (Lawton and Brody, 1969), and a scale designed by the (Re)Programa team called scale of housing conditions-HABITA, which was drawn from architectural indicators in relation to the individual's perception of the characteristics of different rooms inhousings.

- Action Protocol. From the diagnosis of the neighbourhoods studied based on the analysis of detected lesions, energy buildings demand,degree of compliance with accessibility standards and requirements, preferences and expectations of elderly, we develop a comprehensive relationship of complete rehabilitation activities economically valued, including depreciation and cost optimal [7].

We establish three levels of intervention of varying intensity (slight, moderate and severe) in relation to the volume of work, total costs, energy savings, runtime, disturbance to neighbours, need for people relocation, technical, administrative or bureaucratic complexity, the impacts on the environment (dust emissions, noise, vibrations,) and political and social interest. We define an "index of effectiveness" to consider the feasibility of rehabilitation activities, from the characterization of user profiles based on energy consumption considering energy poverty and social reality.

We include as an essential document a report on specific urban regulations, on the changes required and on possibilities of urban intensification that allow works amortization.

CONCLUSION

The architectural refurbishment of neighbourhoods should be undertaken considering the socioeconomic properties of the population with a significant financial vulnerability in a high percentage of cases.

The research project developed by (Re)Programa team at University of Seville, is funded by the Department of Public Works of the Regional Government of Andalusia, Spain, within the PAIDI TEP 954-"InFact" research group, and proposes the design of management and financing tools to feasible rehabilitation operations and urban regeneration with the aim of improving people´s quality of life, avoiding negative impacts on the environment.

Guidelines for action are designed based on the analysis of architectural, urban planning, construction, and social issues, and on the need to respond to regulatory and political requirements, especially regarding energy efficiency aspects and active ageing policies.

(Re)Programa provides tools to quantify energy renovation actions by proposing intervention levels and an index of effectiveness. These tools can be used to assess the viability of any restoration project, adapting to each architectural and social reality. The functional, environmental and spatial adequacy of housing and public transit spaces to cultural, professional, sports or entertainment activities, provide the participation of elderly people in the society, ensuring healthy ageing, and a good quality of life.

It is important to consider simple interventions to reduce energy demand of buildings and encourage mobility, measures approachable from fragile economies and of a high efficiency.

ACKNOWLEDGEMENTS

The authors would like to thank the ERDF of European Union for financial support *via* project "Título del proyecto" of the "Programa Operativo FEDER de Andalucía 2007-2013". We also thank to Public Works Agency and Regional Ministry of Public Works and Housing of the Regional Government of Andalusia.

CONFLICT OF INTEREST

The authors confirm that this chapter contents have no conflict of interest.

REFERENCES

[1] Ministry of Industry, Energy and Tourism of the Government of Spain, "*Document Action Plan and Energy Efficiency Savings* 2011-2020", Madrid, 2008.

[2] Ministry of Public Works of the Government of Spain, "*Law 8/2013 of 26th of June on urban rehabilitation, regeneration and renovation",* Madrid, 2013.

[3] Web site: http://grupo.us.es/reprograma/[March 03, 2015 accessed]

[4] Ministry of Public Works of the Government of Spain, "*State Plan for the Promotion of Rental, Rehabilitation, Urban Regeneration and Renovation 2013–2016*", Madrid, 2013.

[5] Ministry of Public Works of the Government of Spain, "*Technical Building Code*", Madrid, 2009.

[6] M. OlleroBaturone, "*Atención a pacientes pluripatológicos: proceso asistencial integrado*" Seville: Andalusian Health Service of the Andalusian Government, 2007.

[7] European Union, "*Delegated Regulation UE nº 244/2012 of 16th of January 2012 on energy efficiency*", Brussels, 2012.

CHAPTER 7

Energy Efficiency Indicators for Buildings in Argentina

Jorge D. Czajkowski*

Laboratory of Sustainable Architecture and Habitat, Faculty of Architecture and Urbanism, National University of La Plata, CONICET, 47 St. no.162- 1900 La Plata – Buenos Aires Province- Argentina

Abstract: In this work, we make a comparative study of energy performance in different types of buildings. We analyse three cases of office and three of residential buildings, which are compared with samples of previous researches. We look for correlative or different behaviour in terms of potential energy losses and gains. Then, we compare indicators and indexes with Argentinean Standards to verify their efficiency level. We use the "Energo-CAD" methodology and energy model, based on steady-state balances. We discuss the results of energy audits that show discomfort with high energy demand, and their relation to energy breakdown in this country. We conclude that the indicators employed show clear results, and facilitate the comparison among studied cases. There are discrepancies among different size buildings, and we must remark the progressive growth of energy inefficiency. Nevertheless, small buildings are relatively less energy-efficient than large ones. This work ends showing that the sample does not meet the National Standards, even when they are compulsory by law since 2003 in the province of Buenos Aires.

Keywords: Argentina, buildings, comfort level, electricity, energy indicators, energy efficiency, energy audits, energy consumption, energy demand, energy indicators, energy model, energy performance, energy policies, environment, habitat, housing, metropolitan area, natural gas, residential buildings, Standards, sustainable design, thermal load.

INTRODUCTION

This work was developed in the Laboratory of Sustainable Architecture and Habitat – Faculty of Architecture and Urbanism- National University of La Plata (LayHS- FAU- UNLP). It belongs to one of the main research lines developed in our laboratory, and it focuses on building energy efficiency in urban areas, within

***Corresponding author Jorge D. Czajkowski:** Laboratory of Sustainable Architecture and Habitat, Faculty of Architecture and Urbanism, National University of La Plata, CONICET, 47 St. no.162- 1900 La Plata – Buenos Aires Province- Argentina; Lab. Tel: +54 221 4236587/90 Int. 255; Cell: +54 221 5032265; E-mail: jdczajko@gmail.com

Pilar Mercader-Moyano (Ed.)

two framework projects: "Protocol of building construction and sustainabilty indicators for habitat construction° (PIP- Conicet) and "Sustainable buildings certification for Climate Change adaption and mitigation" (UNLP 11/U141).

This issue relates to current problems like lack of resources and global warming, which are— at the same time— tightly connected with building construction and cities [1]. The building industry is one of the most important consumers of raw materials and non-renewable resources. This highly impacts on the environment not only during the extraction and transformation process of raw materials but also during its construction, life cycle, demolition, and recycling [2]. Fossil fuels comprehend the main energy source employed in the built environment. For instance, in Argentina, 96% electricity generates in combinated cycle power plants, while natural gas is intensely used for heating [3].

STUDY AREA CHARACTERISTICS AND SCOPE OF THE ANALYSIS

The scope of this study comprehends buildings located in Buenos Aires Metropolitan Area (BAMA) and Greater La Plata (GLP), together with a few cases situated in Patagonian Extreme South (Tierra del Fuego) and in the Northeast (Misiones).

Argentina´s population —40.117.096 inhabitants— lives in 11,248,194 homes (INDEC, 2010). Now population amounts to 43.131.966 (Jan/2015 estim.) [4].

The main area analyzed (BAMA+GLP) represents 33.64% of the population and 36.12% of the homes. If we considered Buenos Aires City separately from BAMA, 99.6% of the people have sewers, 99.9%, drinking water network, 99%, electricity network, and 65.5%, natural gas network while in Buenos Aires outskirts, 50.3% people have sewers, 75.1% drinking water network, 96.8% electricity network, and 78.4% natural gas network.

The considered area concentrates 33.64% of the Argentine population and consumes 90% of the primary energy for building sub-sector (housing, health, education, administration).

The main sample of the analysed buildings is located in La Plata City. It is situated in IIIb Bioclimatic Zone, whose climate is warm and humid. It is Buenos Aires Province capital city, located 70 km South of Buenos Aires City. With 894,253 people, considering GLP, it is the 5th most populated city in the country.

Despite its specific characteristics, temperate climate in our country stands out because of marked but not extreme summer and winters. High humidity level all year round, with thermal amplitude lower than 14ºC, is its main feature. In the summer, average temperatures oscillate between 20 and 26ºC, with maximum ones over 30ºC. In the winter, average temperatures oscillate between 8 and 12ºC, with minimum ones between 5 and 8ºC. Average relative humidity varies between 70 and 80%. Metropolitan agglomerations collaborate to the "heat island" effect, raising temperatures 3 to 4ºC [5].

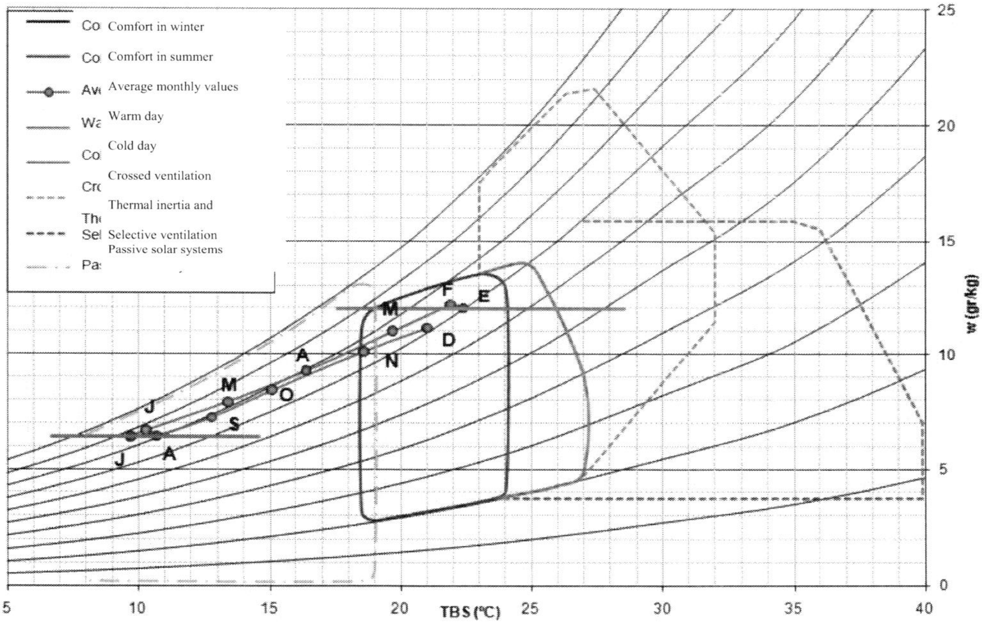

Figure 1: "Givoni climate diagram for La Plata, Argentina". (Source: the author).

For this zone, IRAM Standards recommends light colours, double thermal insulation in roofs respect to walls, ventilated façades in large buildings, double glazing and movable solar protection. It is recommendable to harness dominant winds and protect fenestration looking S and SE because of strong storms. The best orientation is NNW, N, and NNE because it is easier to protect from the sunlight in the summer while receiving sunlight in the winter. According to Givoni´s climate diagram, the use of passive solar systems should be considered most of the year (Fig. **1**).

We found that most of buildings built between 1980 and 2000, had a conventional technique of independent reinforced concrete structure and an external wall made of 12x18x33 hollow bricks without additional thermal insulation. Internal walls are made of 12x18x33 hollow bricks to divide apartments and common areas among each other while 8x18x33 hollow brick walls divide rooms in each apartment.

From 2000 on, regulations authorized 12x18x33 envelopes, diminishing the external wall thickness.

The most common heating system is the balanced draught stove, with 2,300 to 6,000 Kcal/h heat power per room. In 2003, restrictions for new gas network connections increased. At the same time, electricity demand increased as the use of heat/cool split air conditioners, radiant plates and oil radiators massively grew.

In 2012, considering residential and tertiary sectors together (Fig. **2**), building subsector represented 30% of the whole country's energy consumption. 55% of buildings' energy consumption comes from natural gas, 34% from electricity, 9% oil-based liquid fuels (kerosene, LPG, diesel oil), and only 2% from biofuels (firewood).

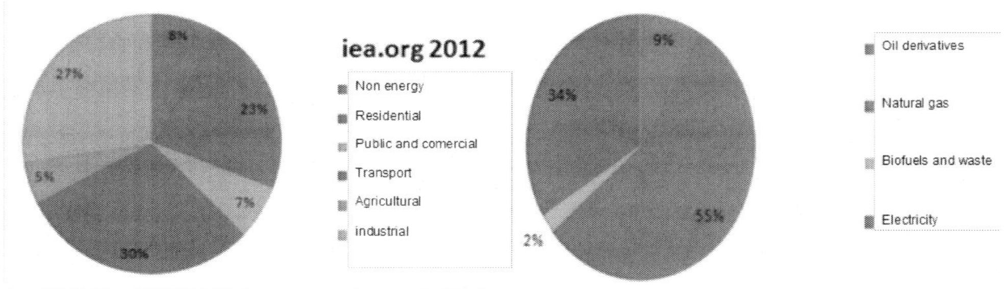

Figure 2: "Argentina Energy end use by sectors and sources (residential + tertiary)" (Source: IEA and the author).

In Fig. **3**, we can see the notable growth of energy consumption in the building sector (residential: 20.3% and tertiary sector: 15.1%) from 2007 to 2013, with the housing sector showing the higher growth. Though transport sector is not in the scope of this work, it shows a remarkable consumption growth (26.7%). In the same period, population grew 5.1%.

A macro analysis from *International Energy Agency* - IEA, shows that in 2007, the residential sector consumed 316.90 Mtoe/inhabitant while in 2012, this value increased to 377.27 Mtoe/inhabitant (16% more). Besides, CO_2 Greenhouse Gas Emissions, which were 4.16 CO_2 Mt/inhab., reached 4.59 CO_2 Mt/inhab. in 2013.

INSTRUMENTS AND METHODS

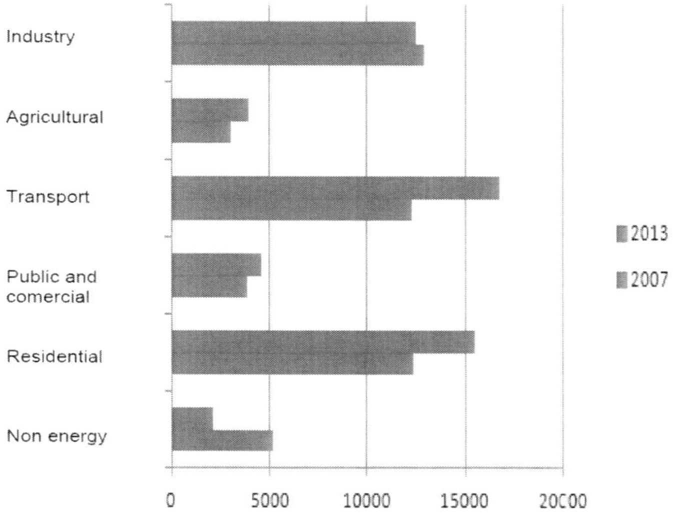

Figure 3: Argentina – "Total energy consumption evolution (secondary) MToe/year". (Source: National Energy Secretary -MECON, author´s elaboration).

This slight 9.32% growth.in residential energy consumption, compared with 20.3% growth supposes a technological change from natural gas to electricity systems for heating.

INSTRUMENTS AND METHODS

For this work, we used social energy surveys, measurements campaigns and previous records, to audit single-family homes and housing apartments for a decade. We employed EnergoCAD [6] and AuditCAD [7] programs to design indicators, indexes, and building analyses. These programs let us analyze monthly energy and thermal building performance, by audited data entry. At the same time, the programs calculate formal indicators based on IRAM Standards [8-12]. Simultaneously, we conducted a survey to determine variations in natural gas and electricity consumption for heating and cooling respectively.

Measurement campaigns were held in summer and winter periods. For them, we used HOBO U10-003 micro data loggers to measure temperature and humidity in inner rooms, and a HOBO ProV2 weather station to measure the same parameters outdoors. A fixed weather station Davis "Vantage Pro2" registered solar radiation. We processed HOBO data with "HOBOwarePro" and "BoxCarPro".

During those periods, we measured natural gas and electricity consumptions in housing units (HU) with four HOBOS in each audited HU. We installed one in the main bedroom, one in the living-dining room, one over the heater and one over the gas cooker. The last ones distinguish the natural gas consumption for sanitary hot water from that for cooking. We fixed data acquisition time every fifteen minutes for HOBOS in the main rooms and the weather station. For HOBOS over the heater and cooker, we fixed it every minute. We exported data to Excel for analysis.

INDICATORS DESIGN

In order to model the built habitat after having analysed its physical, thermal and dimensional characteristics, we crossed variables to build indicators.

One of the most relevant indicators is the relation between heated air volume and annual thermal load (ATL) for HU between 50 and 600m^3.As we show in Fig. **4**, small units are less scattered than bigger ones (Eq. 1).

$$ATL = 53.787 \text{ x Vol} - 2035.7 \quad (1)$$

In Fig. **5**, we can see the relationship between annual thermal load (ATL) in kW/year and the envelope area (walls, floors, windows and roofs).

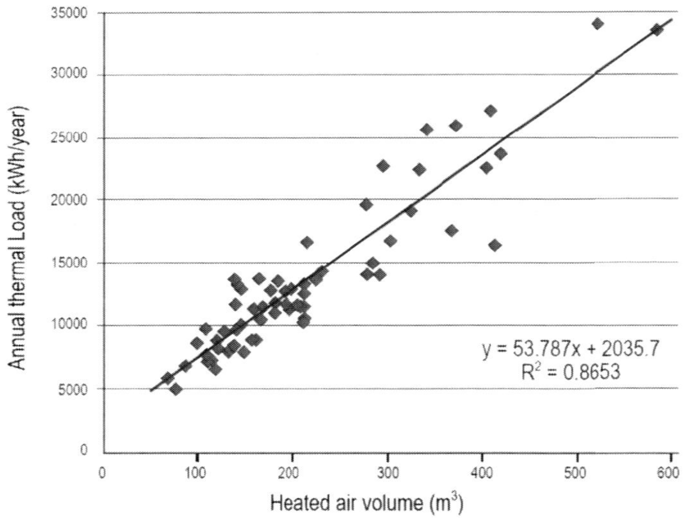

Figure 4: "Annual thermal load/ heated air volume ATL (kWh/year) and heated volume air (m3)". (Source: the author).

A similar situation occurs with R^2= 0.87. This expression let us know the annual thermal load considering HU air volume within a 70-580 m^3 range (Eq. 2).

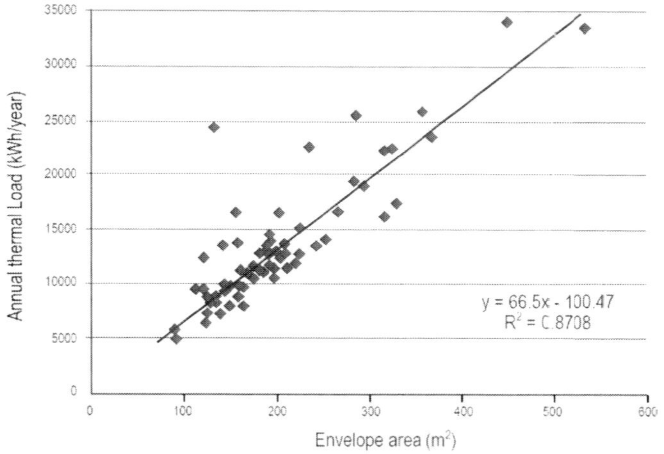

Figure 5: "Annual thermal load/total envelope area- ATL (kWh/year) and total envelope area (m^2)". (Source: the author).

$$ATL = 66.5 \times EnvA - 100.47 \quad (2)$$

We can notice that points in the range from 70 to 230m^3 are less scattered because of higher compactness in HUs, which belong to low/middle to middle/middle socioeconomic sectors.

In cases of air volumes higher than 250m^3, scatter grows due to the combination of compactness reduction and glazing/wall augmentation.

In Fig. **5**, we analyzed the ATL/exposed envelope relationship (Eq. 3). We found a remarkable scatter increase due to urban location characteristics of the audited cases, as we considered high-rise apartment buildings, one-floor apartment blocks, detached houses and others kinds of housing attachments. This situation occurs because the sample represents an urban tissue with high to low density and different degrees of consolidation.

$$ATL = 10648 \times Exp\ A + 4045.2 \quad (3)$$

We can indirectly infer other characteristics from Fig. **6** like the relationship between the distribution in urban space and density. Points situated within 15 to 110m^2 ExpA and 4.900 to 14000 kWh/year ATL correspond to low energy demand Hus, which are located in relatively dense urban tissue. They constitute ¾ of the sample while the remaining ¼ shows a higher exposed envelope area and higher ATL, from 3.500 to 14.000 kWh/year.

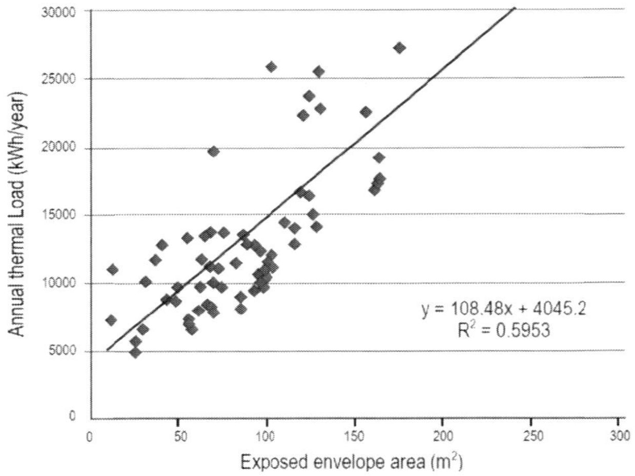

Figure 6: "Annual thermal load/ exposed envelope area ATL (kWh/year) and exposed envelope area (m^2)". (Source: the author).

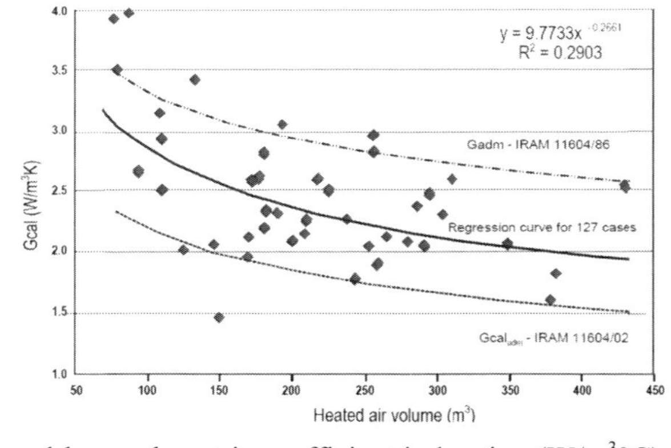

Figure 7: "Gheat thermal loss volumetric coefficient in heating ($W/m^{3}°C$) and heated air volume (m^3) relationship". (Source: the author).

Fig. **7** shows the thermal quality level of the sample, considering the relationship between the volumetric thermal loss coefficient in heating— G_{heat} (W/m^3K)—and the heated volume. G_{heat} is a precise thermal quality indicator for a housing building; the figure shows the heterogeneity of our urban tissue, which reflects our culture. This hinders the definition of policies related to energy saving and rational use of energy, and reduces the possibilities of typifying saving guidelines.

These potencial demand indicators let us work on building thermal quality in our country, energy demand to maintain a certain comfort level, GGE quantification, investment prospects, *etc.*

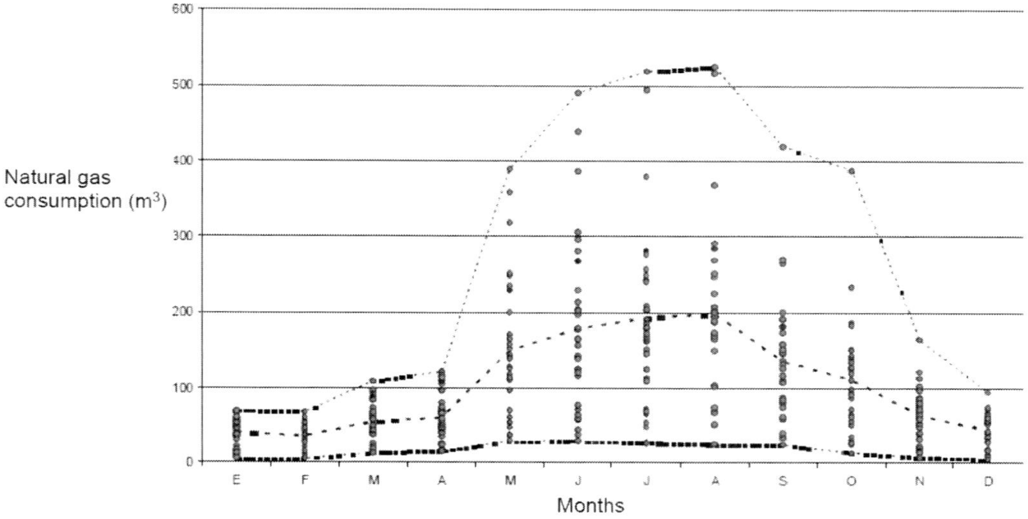

Figure 8: Monthly natural gas consumption distribution (m^3/month) Great La Plata unit housing sample. Maximum, medium and minimum curves". (Source: the authcr).

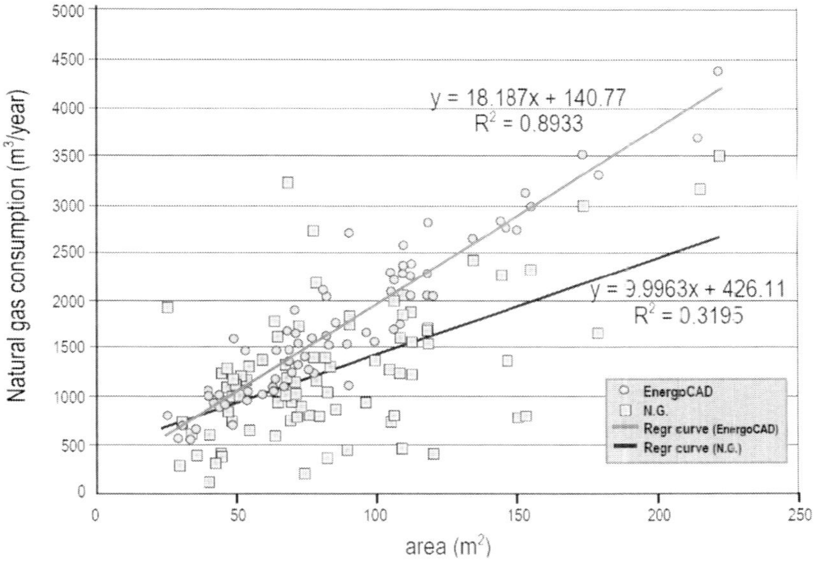

Figure 9: "Comparison between natural gas consumption and demand in heating (BAMA HUs)" (Source: the author).

Other indicators that appear when we collect real energy consumption data let us approach the issue of energy efficiency in buildings. In 2006, the Ente Nacional Regulador del Gas (ENARGAS) reported that 40.9% of Argentina's energy consumption corresponded to sanitary hot water and cooking with the remaining 59%, to heating.

Our sample of audited cases showed that minimum natural gas consumption curve varies between 5.13 and 29.5 m^3/month. Average consumption curve varies between 35 m^3/month and 193 m^3/month. Maximum consumption curve varies between 68 m^3/month and 525 m^3/month. Point concentration near the middle consumption curve lets us infer the characterization of our sample, which mainly comprehends middle and middle-low consumption sectors (Fig. **8**).

If we divide the sample into three segments: units smaller than 50 m^2, units from 51 to 100 m^2 and units larger than 100 m^2, we can analyse the energy behaviour in each segment separately.

We observe that natural gas demand for heating to reach comfortable temperatures is the following:

a. units smaller than 50 m^2: between 540m^3/year (10,8m^3/m^2year) and 1590m^3/year (31,8 m^3/m^2year)

b. units from 51 to 100 m^2: between 700 m^3/year (13.7m^3/m^2year) and 2700 m^3/year (270m^3/m^2year)

c. units larger than100 m^2: between 1700 m^3/year (170m^3/m^2year) and 4390 m^3/year (439m^3/m^2year)

In Fig. **9** we can observe the relationship between consumption and natural gas demand for heating, with a significant dispersion (R^2=0.3195) when we measured consumption. This behaviour is the consequence of different individual heating systems, and variable comfort levels. The *EnergoCAD* calculated demand according to heated area. The results have a high level of confidence (R^2= 0.8933), as shown in Equation 4:

$$AD_{NG} = 18.187 \times A_{heat} + 140.77 \quad (4)$$

AD_{NG}: Natural gas annual potencial demand (m^3/year)

A_{heat}: Housing unit heated area (m^2)

As we mentioned above, real consumption scatter is much higher (R^2=0.3195) and it can be empirically calculated as follows (Eq. 5)

$$CAC_{NG} = 9.9963 \times A_{heat} + 426.11 \quad (5)$$

AC_{NG}: Natural gas annual consumption (m^3/year)

A_{heat}: Housing unit heated area (m^2)

As analysed cases built in different years showed a G_{heat} decreasing tendency, we made a comparative analysis for housing and office buildings (Fig. **10**). The results show that there is an increasing inefficiency over time, with higher incidence on housing than on office buildings, which demonstrates the lack of an effective regulation on building thermal quality.

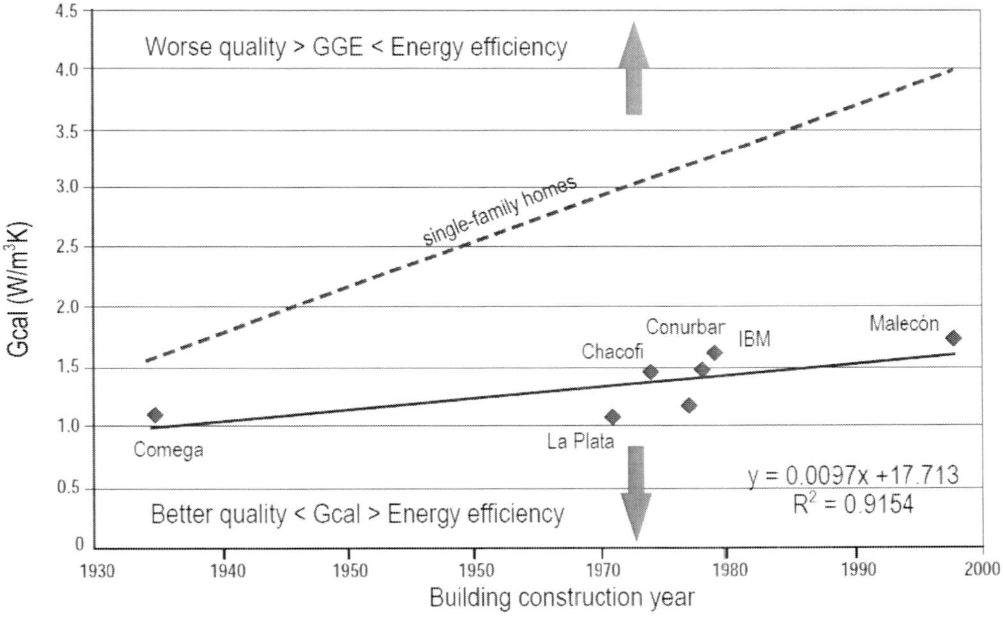

Figure 10: Comparation between the energy unefficiency level of housing and office buildings in Argentina between 1900-2000. (Source: the author).

This absence of regulations leads to particular situations as hygrothermal discomfort in urban middle-high social sectors. Despite different facts ranging from huge subsidies— that ruled until 2014— to final energy consumption in natural gas and electricity, residential hygrothermal discomfort is significantly high (Fig. **11**).

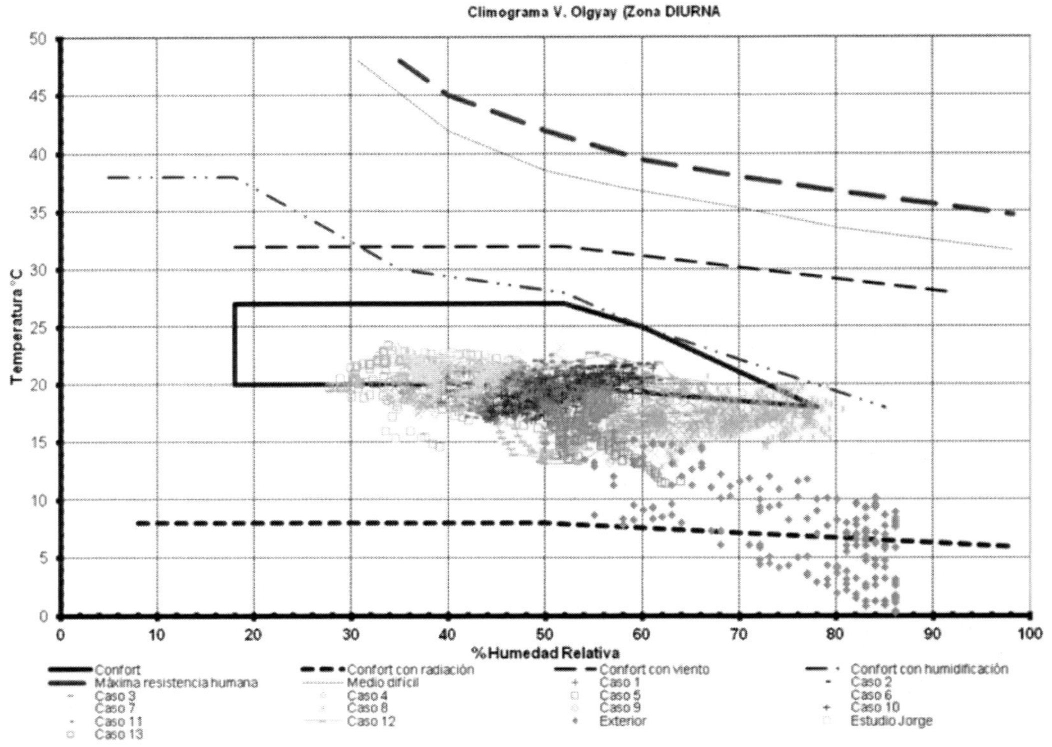

Figure 11: Hygrothermal discomfort: middle high-income HUs in La Plata". Source: the author.

Natural Gas Consumption in Argentina and Other Countries

We mentioned above that 25.5% of the natural gas generated electricity for combined cycle thermal power plants, and 32% supplied the building sector.

ENARGAS estimates that 13% corresponds to a basic average annual consumption for sanitary hot water and cooking, while 20.5% corresponds to heating.

We conduct this analysis in order to compare energy demand for heating in buildings, being aware that it is not a simple task due to the great scatter in building codes, construction systems, ways of energy use, rigour of climate, among others.

Currently, the most used indicator measures total energy demand (heating, cooling, lighting, equipment, cooking and sanitary hot water) in kWh/year or kWh/m^{2}year.

For statistical purposes, these units of measurement are considered separately, but most authors [12-16], employ them, circumscribed to their own case, country or region. This hinders comparisons among different countries.

One way consists of removing the climatic factor, and dividing annual energy demand by heating degree days (comfort temp. = 18°C), which is a climate rigour indicator for a certain location. In this way, we can calculate standards from different countries [17]. In Europe, Directive 2002/91/EC proposes that each country establishes its own standards, and makes them compulsory by means of Building Codes, leaving its enforcement to each municipality. In Argentina, despite of having IRAM Standards to rule building thermal quality, they are not compulsory.

Buenos Aires Province promulgated Law 13059/03 together with its reglamentary decreet 1030/10, which made compulsory Building Thermal Conditioning IRAM Standards, delegating to municipalities the control of enforcement. But they do not require its enforcement or just delegate the responsability on the profesional. Buenos Aires City promulgated Law 4458/12, with the addition of energy efficiency labelling in heating and cooling but does not enforce it.

In Rosario City, the local government added energy efficiency regulations to its Building Code by means of Ordinance 8757/11. Since it is not compulsory, construction market together with professional colleges has the true control to maximize real state investors benefit.

In order to know energy state in the built habitat, we propose an analysis to find energy consumption indicators in buildings and compare them with national regulators data and international values. Then, we can quantify thermal quality state in buildings in Argentina.

This analysis is divided in two sections: natural gas consumptions and potential demand to reach thermal confort (18°C), collecting data from a sample of 500 audited HU cases.

We begin by calculating an energy quality indicator, G_{heat}, in W/m^3K[12]. This indicator determines building thermal quality and other parameters. We can calculate natural gas consumption in heating, using temperature difference degree by the following equation (Eq. 6):

$$G_{heat} \text{ x } V \text{ x } T_{heat} \text{ x } HHF$$

HC°C = ----------------------------- (6)

Hp x p

HC°C: Natural gas consumption per day per temperature difference degree [m^3/°Cday]; G_{heat}: Heat loss volumetric coefficient [W/m^3K]; V: Heated indoor air volume [m^3]; T_{heat}: Heating average time [h/day]; HHF: Heating housing factor: % [adim]; Hc: Natural gas heat power [W/m^3] and p: Heating system average performance [adim].

We group audited cases in three segments called: *compact house, apartment* and *large house*.

Table 1: HU types and morphological and thermal characteristics

	Gm_{heat} W/m^3°C	Vm m3	Am m^2	Weight %	Theat	AHF
Compact house	2.98	180	67	60	12	0.45
Apartment	2.5	100	37	30	12	0.7
Large house	2.15	350	130	10	24	0.7

Table **1** shows average values summary for Gm_{heat}, average heated volume *Vm*(m^3), average heated area *Am* (m^2), percentage rate respect to urban tissue in middle/large cities, average heating time according to surveys, *Theat* (h/day) and average heating percentage factor (%)

We determine specific natural gas consumptions for the three types of HUs, compact type: 1.2m^3/°C day, apartment type: 0.56m^3/°C day, and large type: 2.36m^3/°C day. Using weights from Table **1**, we can calculate weighted specific natural gas consumption—$HC_{weighted}$, according to the following equation (Eq. 7):

$$HC_{weighted} = 0.54 \times 0.6 + 0.56 \times 0.3 + 2.36 \times 0.1 = 0.73 \text{ m}^3/°\text{C day} \quad (7)$$

We obtain the value 0.73m^3/°C day by auditing housing characteristics and making calculations from average values. We noticed that— regardless the building type—a considerable percentage of people barely heat their homes.

We are not taking into account solar gain effect, which represents 12% to 30% of sunny winter days. This consideration would reduce energy consumption at a certain amount.

The obtained mean weighted value differs from ENARGAS value, which is 56 m^3/°C day. Since *ENARGAS* average value is 30% lower than ours— which was obtained by means of audits—, it could imply that people do not heat their homes as much as they declare in surveys. However, it could happen that the considered sample is biased because of showing middle/middle-high social sectors behaviour. Thus, the impact of low-income sector on the results is considerable, even when they live in homes with similar dimensions and construction systems.

As a feasible hypothesis, we infer that they heat smaller areas, during much less time, achieving inner temperatures lower than the comfort level (18°C).

In Argentina, though there are standards to regulate building thermal quality, in practice, they are not compulsory neither in private nor in public sectors.

As shown above, we can display three different scenarios with average annual demand per HU in BAMA, connected to natural gas network:

- *Argentina a:* the energy demand per HU, within thermal comfort range (18 to 20°C) during 24 hours a day in winter and with 90% of HU heated area.
- *Argentina b:* Real energy demand that results from energy audits along 20 years in the region, with a partially heated HU and inner temperature below comfort level.
- *Argentina IRAM 11604:* Energy demand according to "Saving energy in heating" model, presented below.

In Fig. **12**, compare homes in different countries. We can clearly demonstrate that an average HU— which represents more than 3.62 millions in BAMA— is 28% less efficient (0.32 kWh/m^2 °HDD year) than a Spanish average HU (0.23 kWh/m^2°HDD year) and 93% less efficient than a German HU (0.023 kWh/m^2°HDD year). We consider average HU built according to current Building Codes for each location.

As we do not know the requisites for housing thermal quality apart from official plans, we can assume *Argentina a*, as an average national value.

CONCLUSION

Energy indicators employed by national or provincial governments in Argentina are calculated with data provided by energy companies: energy offer and national statistics about population and housing. These indicators never result from energy demand and do not let us infer energy use to achieve comfort level. We are not able to know if there is a potential reservoir of energy saving and rational use of energy, nor if this saving is feasible. Indicators proposed in this work try to obtain better information based on energy demand and hygrothermal comfort satisfaction degree. Indicators are built considering audited cases that are occupied now.

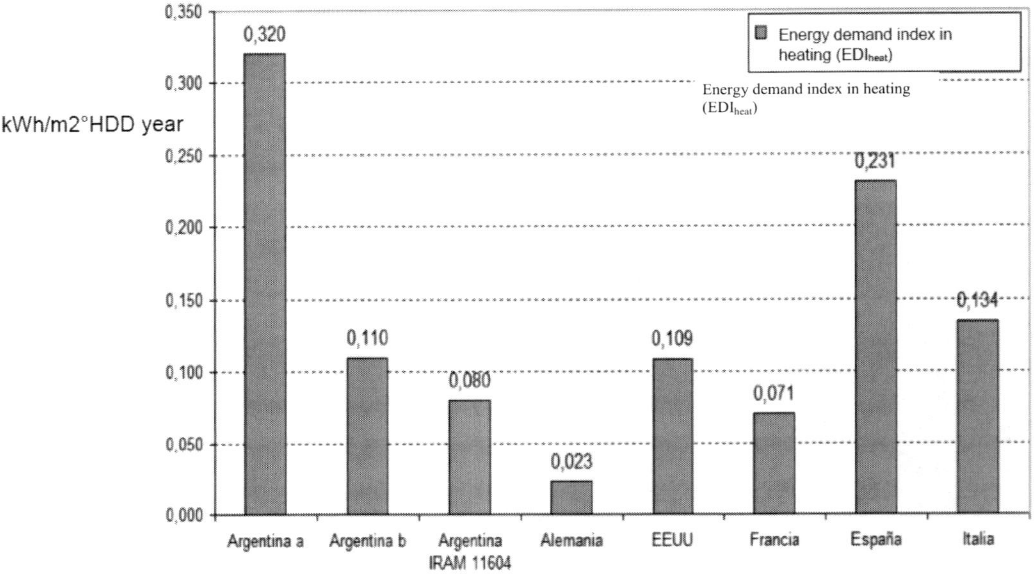

Figure 12: "Comparison between energy demand index for heating (EDI_{heat}), in medium-sized Hus, considering three feasible scenarios in Argentina and other countries. Source: the author.

Energy efficiency in buildings, such as rules, laws, or building codes, are not enough regulation instruments if national, provincial or municipal policies tend to deregulate. This lack of energy policies to rule demand provokes economic and energy breakdowns, perpetual energy crisis, absence of construction innovation, GGE growth, among other consequences. These are all issues that a developing country faces.

ACKNOWLEDGEMENTS

A especial acknowledgement to my team members, with whom go forward in the knowledge of BAMA built habitat: Analía Gómez, Cecilia Corredera, Carolina Vagge, Belén Salvetti, Mariela Marcilese, Paz Diulio and Mauro García Santa Cruz.

To the organizations which financed our research: National University of La Plata, National Council of Scientific and Technological Research and National Agency for Scientific and Technological Promotion.

CONFLICT OF INTEREST

The author confirms that this chapter contents have no conflict of interest.

REFERENCES

[1] IPCC, "*Third Evaluation Report-Intergovernmental Panel on Climate Change*", Geneve, Switzerland, 2001. ISBN 0 52180770 0.

[2] B. Edwards, "*Guía básica de la sostenibilidad*".Barcelona: Gustavo Gili, SL, 2008. ISBN 8425222087.

[3] Nationl Energy Secretary, *National Energy Balance 2013*. National Ministery of Economy, Buenos Aires, 2014. [Online]: http://www.energia.gov.ar/. /balances_2014/Ben13.xls

[4] INDEC (2015). Estimaciones y proyecciones de población 2010-2040. Total del país.

[5] J. Czajkowski, A. Gómez, "*Arquitectura Sustentable. Arq. Clarín*", Buenos Aires, 2009. ISBN 978-987-07-0603-8.

[6] Czajkowski, Corredera, Saposnik, "*A*nálisis de la relación entre demanda de gas natural en calefacción según "*EnergoCAD*" y consumos reales en viviendas unifamiliares del Gran La Plata", *Revista Avances en energías renovables y ambientales*, vol. 7, 2003. ISSN 0329-5184.

[7] J. Czajkowski, "*Desarrollo del programa AuditCAD para el análisis de edificios a partir de auditorías ambientales*", *Revista Avances en energías renovables y ambientales*, vol.3, 1993. ISSN 0329-5184.

[8] IRAM. 11659-2, "*Building thermal insulation Hygrothermal conditions verification. Saving energy in cooling Parte2: Housing units*", Certification and Standardization Argentine Institute Buenos Aires, 2007.

[9] IRAM. 11601, "*Buildng thermal insulation Calculation methods IRAM*", Buenos Aires, 2002.

[10] IRAM. 11603, "*Building thermal conditioning. Argentine Republic Bioclimatic Clasiffication*", Certification and Standardization Argentine Institute, Buenos Aires, 1996.

[11] IRAM. 11605, "*Building thermal conditioning. Building habitability conditions. IRAM*", Buenos Aires. 1996.

[12] IRAM. 11604, "*Building thermal insulation Hygrothermal conditions verification. Saving energy in heating. G Heat loss volumetric coefficient Calculation and limit values*", Certification and Standardization Argentine Institute, Buenos Aires, 1990.

[13] F. Asdrubali, M. Bonaut, M. Battisti, M. Venegas, "Comparative study of energy regulations for buildings in Italy and Spain", *Energy and Buildings* vol. 40, pp.1805–1815, 2008.

[14] T. Maier, M. Krazaczrk, J. Tejchman, "Comparison of physical performances of the ventilation systems in low-energy residential houses", *Energy and Buildings* vol. 41, pp. 337–353, 2009.

[15] C. Tiberiu, J. Virgone, E. Blanco, "Development and validation of regression models to predict monthly heating demand for residential buildings", *Energy and Buildings* vol. 40, pp. 1825–1832, 2008.

[16] S. Heiple, J. Sailor David, "Using building energy simulation and geospatial modeling techniques to determine high resolution building sector energy consumption profiles", *Energy and Buildings* vol. 40, pp. 1426–1436, 2008.

[17] PREDAC. Decreto 192; 2005. CSTB, 2005. CTE, 2006.

Chapter 8: Towards A Sustainable Retrofitting Plan for Social Housing in Mediterranean Europe

CHAPTER 8

Towards A Sustainable Retrofitting Plan for Social Housing in Mediterranean Europe

Mónica Martínez-Hervas, Juan J. Sendra* **and Rafael Suárez**

Instituto Universitario de Arquitectura y Ciencias de la Construcción, Escuela Técnica Superior de Arquitectura, Universidad de Sevilla, Reina Mercedes 2 Avenue, 41012, Sevilla, Spain

Abstract: The new legal approach proposed by European Directive 2012/27/EU makes it necessary to adapt Spanish regulations and policies of the Member States to the new demands. In order to reach the goals for the reduction of energy consumption and the improvement of energy efficiency in buildings, it is necessary to act on the existing housing stock, which is mostly obsolete, and to encourage energy retrofitting. In this context, research is focused on creating an energy assessment methodology on an urban scale and applying this to a characteristically Mediterranean city in the south of Spain, Cadiz. This research aims to cover the lack of information on the energy behaviour of the existing housing stock, particularly in social housing neighbourhoods built from 1940 to 1980, after the Spanish Civil War and before the building regulations which resulted from the 1970s energy crisis, as a step prior to the establishment of a Plan for Sustainable Retrofitting of social housing in the city of Cadiz. In order to implement this plan, it was necessary to identify and classify the main social housing neighbourhoods built in that period, offering a construction and energy analysis of the constructive systems used in the thermal envelope, which has made it possible to classify and establish their degree of obsolescence. This study has provided a global assessment of the energy rating of social housing built in the city in 1940-1980, applied to a map of the city allowing us to view an energy rating map of sorts of the city. This method is essential in order to attain what could be described as an essential first phase towards attaining the priority action goals of Horizon 2020, using global actions with sustainable retrofitting plans.

Keywords: Energy characterisation, energy rating, Mediterranean city, social housing, sustainable retrofitting.

INTRODUCTION

In the post-war period of the second half of the 20th century, European cities experienced major urban growth in comparison to their historic growth. Given that this housing was built in the pre-regulation period, these houses lack thermal

***Corresponding author Juan J. Sendra:** Instituto Universitario de Arquitectura y Ciencias de la Construcción, Higher Technical School of Architecture, University of Seville, Seville, Spain; Tel: +34 954551630 / +34 639113128; Fax: +34 9544557024; E-mail: jsendra@us.es

insulation measures in their envelope. At present, residential building stock accounts for 40% of European energy consumption [1]. Approximately 40% of this housing stock was built before 1960 [2]. In Mediterranean Europe, with milder climates than those of central or northern Europe, social housing built between 1940 and 1980 presents major thermal deficiencies in envelopes. The resulting major energy losses in turn require high energy consumption to reach the current operating conditions stipulated for housing in the regulations. This all leads to very low energy rating.

Therefore, there is extensive residential building stock in European Mediterranean cities with a major potential for energy saving, prompting the implementation of strategies and policies aimed at the improvement of energy efficiency [3]. This situation brought about the appearance of new regulations and policies among the strategies of Horizon 2020, such as Directive 2012/27/EU, which states that "Member States shall establish a long-term strategy for mobilising investment in the renovation of the national stock of residential and commercial buildings, both public and private". These national long-term strategies must be based on new global action policies, mostly through large-scale energy retrofitting, which requires an appropriate diagnosis and knowledge of the residential building stock, in order to identify and optimise the potential for improving energy efficiency through energy retrofitting processes.

Up until now extensive research has been carried out mainly in the climate zones of central and northern Europe to provide an energy characterisation of housing stock and analyse the economic viability of its retrofitting following energy criteria [4, 5]. However, there are notable variations in its application in the climate zones of the southern Mediterranean and, moreover, assessment methods followed do not always correspond to the energy behaviour expected [6]. To do this, new tools for assessing potential energy improvement were needed, taking into account both residential type [7] and urban scale [8].

In many cases there is a need to speed up the process for energy assessment of buildings in cities [9], as even if they do not offer comprehensive results on energy efficiency they provide information which can be used by the different administrations to plan retrofitting strategies on an urban scale [10, 11]. Some studies focusing on this are to be found among member states of Mediterranean Europe, including Italy, Greece and Cyprus [12-15].

Many of the studies carried out to date follow a methodology that estimates energy consumption, applying different modelling techniques which are included in the bibliographical review by Swan *et al.* [16].

This work aims to offer a methodology to examine the potential for improvement and energy saving on an urban scale, applying it to a characteristic climate zone of southern Spain, and studying the city of Cadiz. The main aims are the identification, characterisation, assessment and energy rating of social housing built between 1940 and 1980, a period predating the first regulations on thermal conditions in Spanish buildings, through the energy analysis of a broad sample of 7372 housing units, 48% of the social housing built in the period of reference. This is an essential first step towards drawing up a Plan for Sustainable Retrofitting of social housing in the city.

METHODOLOGY

The southern Mediterranean Spanish city of Cadiz was chosen as a case study to develop the methodology. The city has mild winters from December to March, warm summers from June to September and mean relative humidity ranging between 68 and 76% (Table **1**). This city is located in climate zone A3 in Spain [17].

Table 1: Climate data for Cadiz, 1981-2010, taken from the Abridged guide to climate in Spain. Source: State Meteorology Agency, Directorate of Production and Infrastructures

Month	Mean temperature (°C)	Maximum mean temperature (°C)	Minimum mean temperature (°C)	Mean rainfall (mm)	Relative humidity (%)
January	12.7	14.8	11.2	69.0	75
February	13.8	15.9	11.2	58.5	74
March	15.5	17.5	13.5	34.7	71
April	16.8	19.2	14.2	45.2	69
May	19.1	20.9	16.3	26.9	70
June	22.4	24.4	19.1	6.7	69
July	24.6	26.8	21.9	0.2	68
August	25.0	27.8	22.8	1.7	70
September	23.3	25.5	20.4	23.8	71
October	20.3	22.3	17.7	67.3	74
November	16.5	19.0	14.8	97.7	74
December	13.9	15.9	22.8	92.3	76
Year	18.6	19.1	17.6	522.8	

A methodology analysing application to any city or climate zone was used to carry out this study, based on the original data for the social housing projects of the study sample, in order to characterise and qualify the original energy condition of a large part of the city of Cadiz. This consisted of three phases: the typological and constructive identification and characterisation of social housing built between 1940 and 1980, the energy assessment of its original condition, and finally the creation of an energy rating map for the city.

Typological and Constructive Identification and Characterisation

The first phase of the work focused on identifying and analysing the type and constructive features of the social housing neighbourhoods in Cadiz, as well as the housing built at different stages within each neighbourhood in the 1940-1980 period. All the social housing developments were identified by studying the historical evolution of Cadiz and its urban plans, as well as consulting different monographs, the Electronic Site of the General Directorate for Cadastre [18] and carrying out a superficial visual inspection using Google Maps. This identification was followed by an initial database of the social housing developments, including information relating to location, developer company or organisation, year of construction, plot surface area, occupied surface area, built surface area, type of building and documentation available. None of this information is available at present and it is essential for establishing the energy status of social housing, as well as being potentially extremely useful for future research.

An analysis was subsequently carried out of the largest neighbourhoods and developments in the city of Cadiz. Different criteria were followed to select these social housing neighbourhoods and developments:

- Representation of the developments within the urban surroundings of the city.
- Number of housing units in the developments.
- Access to sources of information on the original projects for the social housing developments.

The phases selected according to these criteria were then used to create a second database, which expanded on the first with information on the general situation and data of the housing: year of project and construction, social housing property regimes, architect and construction company, number of blocks and housing units and different types of surface. The building and housing unit types and their supporting structures were also defined, and information was compiled on the thermal envelope of the façades, including openings and their solar protection, and roofs. Equally, a brief description was included on the state of conservation and any modifications of the original project which can be observed. Finally, the documentation and information sources available were described.

Before creating this database it was necessary to document the original projects of each of the developments analysed, consulting the Provincial Historical Archives and the Municipal Historical Archives in Cadiz.

The information gathered was used to create technical files on each of the social housing developments, putting together a catalogue which includes the main aspects of the architectural configuration -typology, number of storeys, programme, usable surface area and free height- and of the construction systems - façade, roof and openings (Fig. **1**).

This information enabled us to characterise the architecture of social housing in the city, through statistical analysis of the configuration of the architecture and the construction systems used in each of the periods analysed.

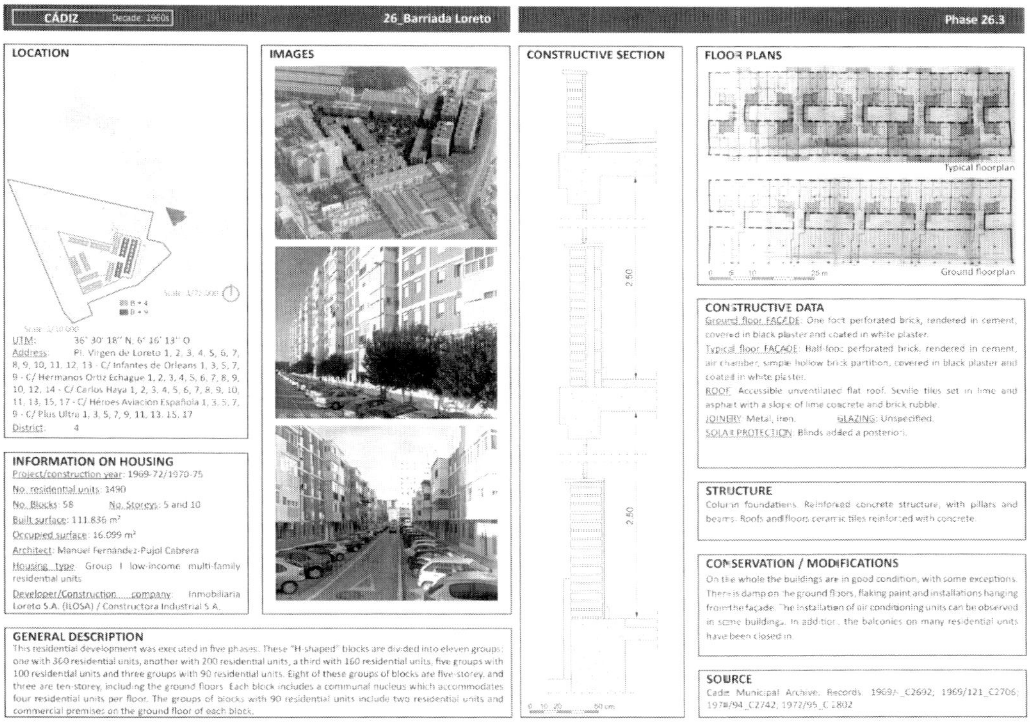

Figure 1: "Technical analysis file. 1490 housing units, Loreto Neighbourhood in Cadiz". (Source: authors).

Energy Assessment

In a second phase, energy assessments were carried out on the original conditions of the social housing developments. Energy models were built for each of the housing developments selected and in general two models were distinguished:

- Housing model, for the simulation of the energy behaviour of the individual housing unit, for all orientations and in three possible situations: ground or first floor housing unit (in the case of commercial premises on the ground floor), housing unit on intermediate floors and housing units below the roof on the top floor.

- Housing block model, for the simulation of the energy behaviour of the entire block, made up of all the communication nuclei together with the buildings they provide access to. Energy models were produced for every orientation and three possible positions - corner, central and isolated - were established.

The energy models were built using the CE^3X calculation program, recognised by the Spanish Government. This program follows a simplified calculation procedure for existing housing [19]. The construction of these models requires prior definition of location, climate data and conditions for use and operation following national regulations [20].

Subsequently, an energy characterisation of the envelope of each energy model was carried out, specifying the thermal transmittances and masses of each of the enclosures which make up the thermal envelope, and the characteristics of the openings, joinery and solar protection. Finally, given that the housing units did not have any active thermal conditioning systems, the systems of reference were included in the Spanish regulations [20].

Following the construction of the energy models, energy assessments were carried out on both the individual housing units and the residential blocks, based on the results for heating, cooling and global demand, as well as those for energy rating. These results were recorded in technical files for each type of construction within each housing development (Fig. **2**).

Energy Rating Map

Finally, a map was created showing the original energy ratings of the social housing neighbourhoods in the city of Cadiz. This map defined the original conditions of the energy situation of the social housing neighbourhoods in the city, essential information for future actions in energy planning and identifying strategies for energy retrofitting.

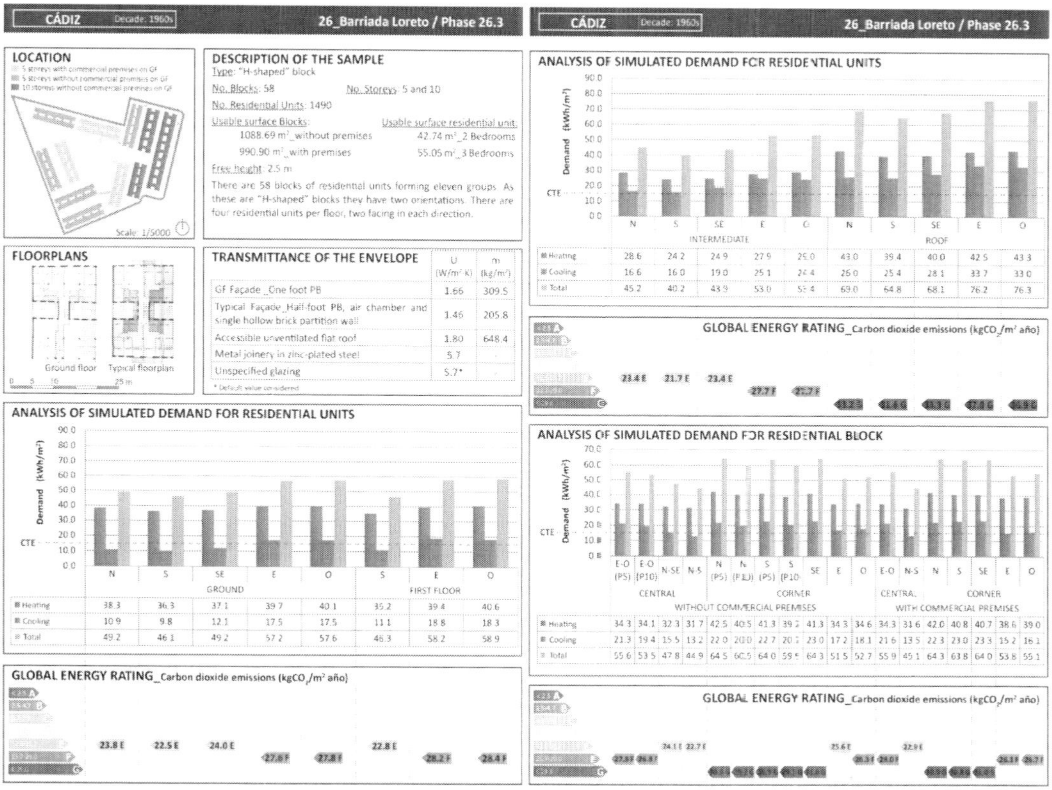

Figure 2: "Technical file for results. 1490 housing units, Loreto Neighbourhood in Cadiz". (Source: authors).

RESULT ANALYSIS

Based on the previous methodology, developed in three phases, the results obtained for each of the different tasks developed were analysed.

Identification of Social Housing Neighbourhoods and Developments

A total of 72 social housing developments, that is 15,233 housing units, were identified in Cadiz. Following the selection criteria, 23 developments were selected for analysis, making up around 50% of the housing units identified (Fig. **3**), and offering a reasonably faithful illustration of the original energy conditions of the housing built in Cadiz in the period analysed.

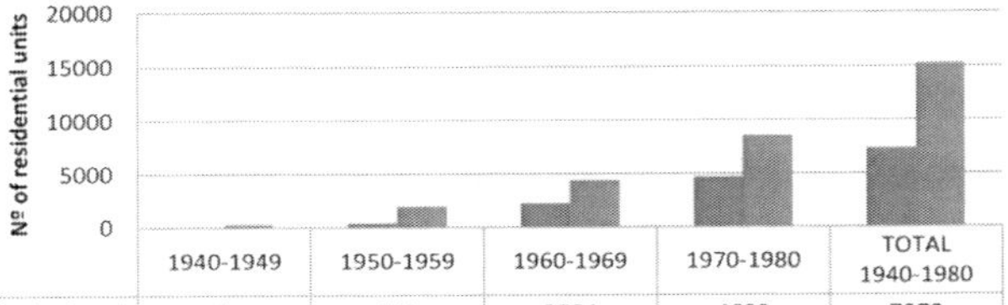

	1940-1949	1950-1959	1960-1969	1970-1980	TOTAL 1940-1980
Analysed	0	449	2224	4699	7372
Identified	315	1970	4389	8559	15233

Figure 3: "Number of social housing units identified and analysed in Cadiz (1940-80)". Source: authors.

Statistical Analysis of the Architectural Configuration

Firstly, in order to characterise the different types of residential blocks their building typology was identified, and a classification was carried out for subsequent analyses. It was observed that throughout the period analysed there were two predominant architectural typologies: linear block and "H-shaped" block. The linear block type was clearly predominant in the first half of the period analysed, while in the 1960s it alternated with the "H-shaped" block type, which would become the predominant type in the 1970s (Fig. **4**). When typological analysis is complemented with an analysis of the number of storeys, an increase can be seen in the number of blocks throughout the period analysed, although blocks never exceed 14 floors (Fig. **5**).

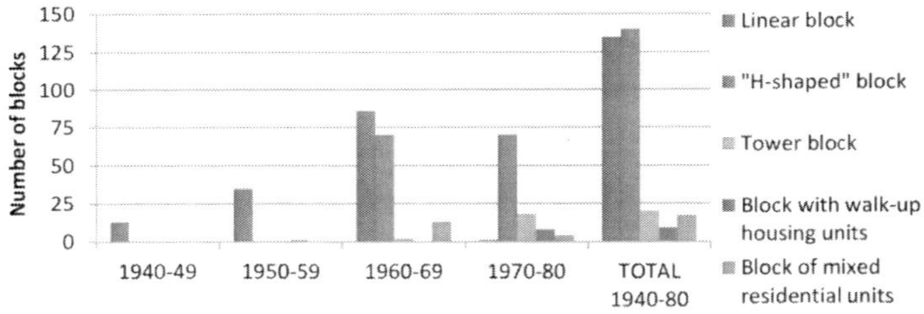

Figure 4: "Evolution of the number of blocks for each typology". (Source: authors).

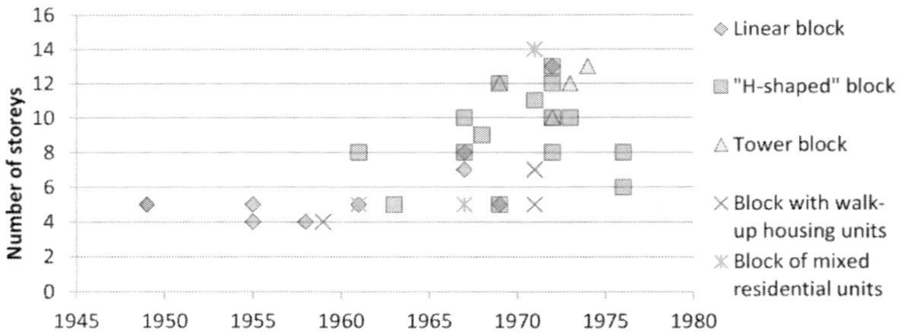

Figure 5: "Evolution of the number of storeys for each building typology". (Source: authors).

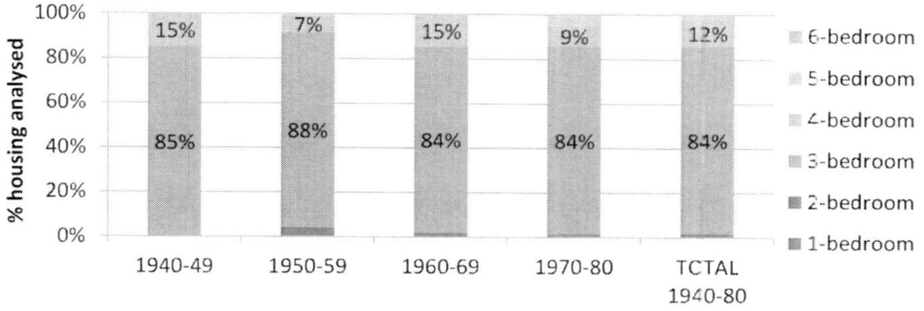

Figure 6: "Evolution of the percentages of the number of residential units analysed according to the number of bedrooms". (Source: authors).

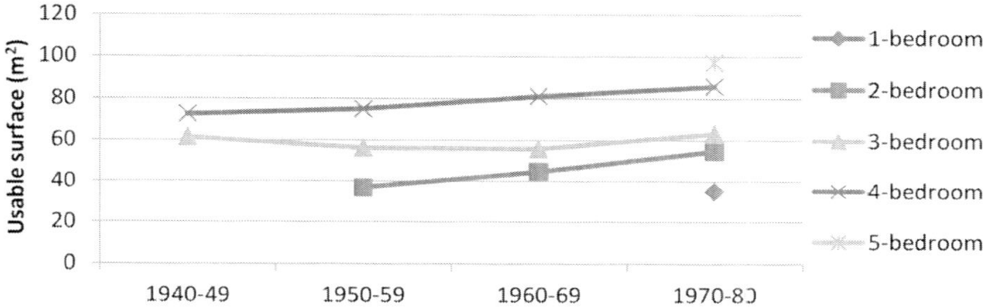

Figure 7: "Evolution of the average usable surface according to the number of bedrooms". (Source: authors).

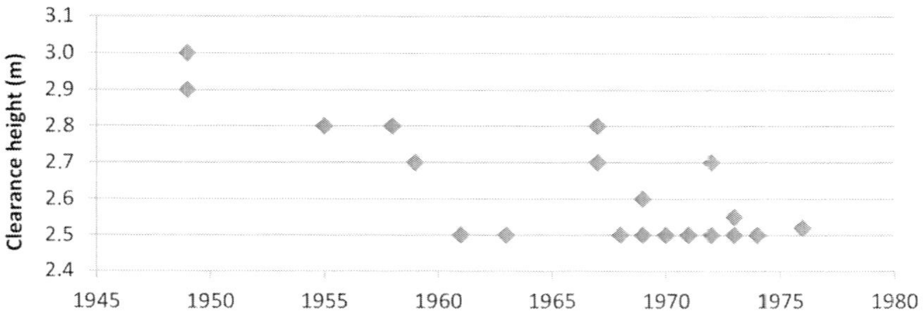

Figure 8: "Evolution of the clearance height of residential units". (Source: authors).

As regards housing units, it was observed that during the four decades studied there was a predominance of three-bedroom homes (Fig. **6**). The usable surface area of these residential units increased slightly over the years, except in the case of three-bedroom residential units, which decreased in the first half of the period studied, but increased in the second half (Fig. **7**). Over the years the clearance height of the residential units decreased from 3 m in the first stage to 2.5 m in the 1970s (Fig. **8**).

Statistical Analysis of Construction Systems Used

Based on the statistical analysis of the constructive characterisation of all the social housing developments analysed throughout the period of study, the most frequently used constructive solutions for the envelope were obtained.

On the façade there is a predominance of the constructive solution using double layer of half-foot solid or perforated bricks, air chamber and partition wall (Fig. **9**), although it was only found in the last two decades. Single-layer solutions are in the minority over the whole period examined, although these can be considered representative types in the early decades. As regards building solutions used on the roof, there is an abundance of unventilated flat roofs (Fig. **10**). The openings mostly incorporate metal frames (Fig. **11**) and 3 mm or 3+3 mm single glazing, although in the projects for many developments the type of glazing was not specified (Fig. **12**). The most common solar protections were PVC blinds although there was also a high percentage of openings with no solar protection (Fig. **13**).

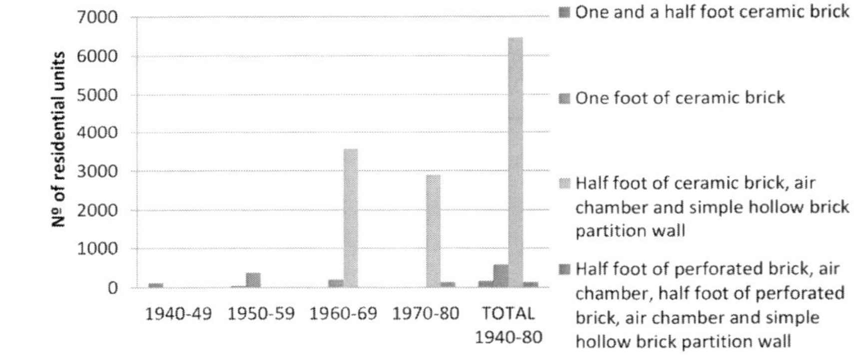

Figure 9: "Evolution of the number of residential units analysed according to the constructive system of the façade". (Source: authors).

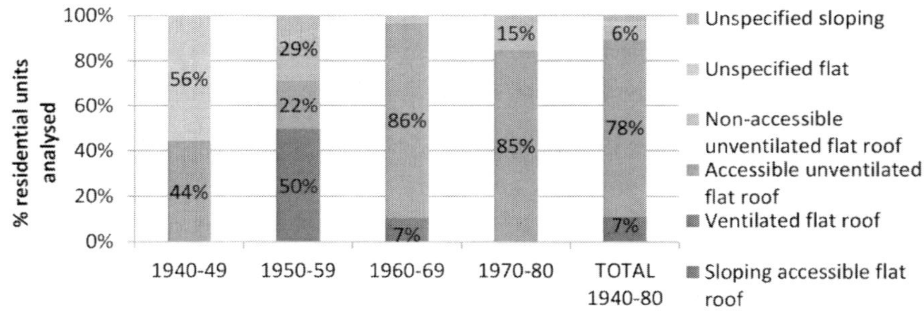

Figure 10: "Evolution of the percentage of residential units analysed according to the constructive system of the roof". (Source: authors).

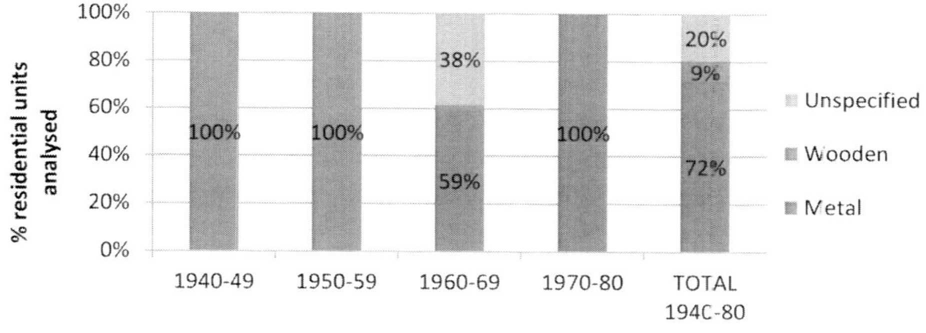

Figure 11: "Evolution of the percentage of residential units analysed according to the joinery used in the openings". (Source: authors).

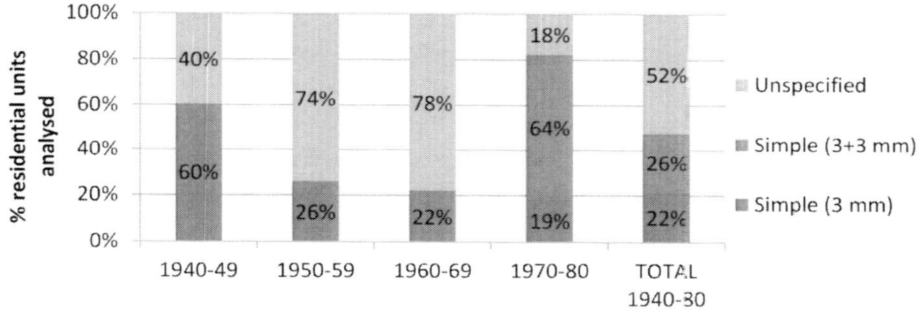

Figure 12: "Evolution of the percentage of residential units analysed according to the type of glazing used in openings". (Source: authors).

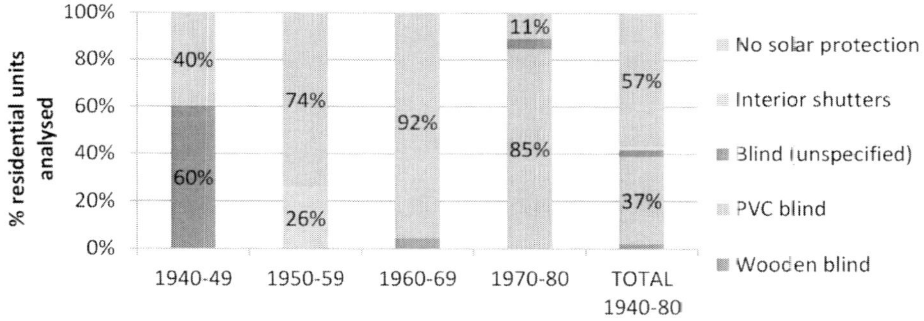

Figure 13: "Evolution of the percentage of residential units analysed according to the solar protection used in openings". (Source: authors).

Energy Characterisation of the Envelope and Construction Systems

Following the analysis of the constructive systems, the thermal envelope, which had high thermal transmittance values due to the lack of thermal insulation, was characterised and analysed. The transmittance values on the façade were between 1.35 and 2.63 W/m^2•K for single-layer solutions, and between 1.37 and 1.63 W/m^2•K for double-layer solutions. These values are far from the demands of current regulations which limit the maximum to 0.94 W/m^2•K. Thermal transmittance values between the 1950s and mid 1960s were higher than those in the 1940s due to the thinner enclosures, all of which were single-layer. However, from the 1960s on, thermal transmittance values fell noticeably due to the generalised use of double brick façades with intermediate air chamber (Fig. **14**).

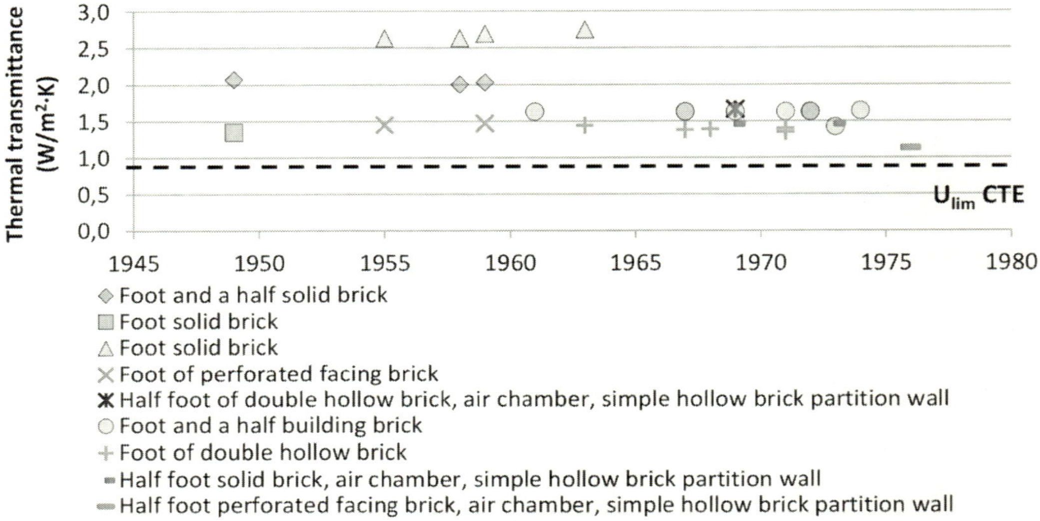

Figure 14: "Evolution of thermal transmittances in constructive solutions for façades". (Source: authors).

As regards the energy characterisation of the constructive solutions in the roof, in the period analysed we found stable transmittance values between 1.5 and 2 W/m^2•K, depending on composition, with variations due to the construction of the slope, except in the case of the ventilated flat roof, which provides lower transmittance values, around 0.86 W/m^2•K, closer to the limit reference value of climate zone A3, 0.5 W/m^2•K (Fig. **15**). In most openings, the predominant metal joinery with single glazing has a transmittance value of 5.7 W/m^2•K.

Energy Assessment of the Housing Unit and Housing Block Models

In the energy assessment of the housing unit model the mean demand values obtained were higher than those stated in the current regulations for climate zone

A3, 15 kWh/m^2. Equally, the values for mean cooling value demands obtained in the residential units exceeded the limit stated in current legislation, also 15 kWh/m^2, although in this case the margin of variation is smaller (Table **2**). It should be noted that housing with a greater envelope surface due to being under a roof presents a heating demand which is on average 60% higher than that of a housing unit in an intermediate position.

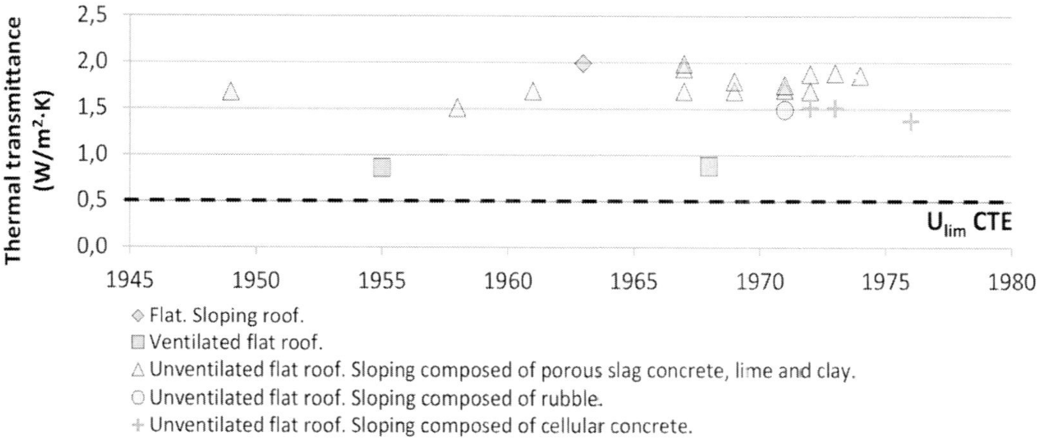

Figure 15: "Evolution of the thermal transmittances of the constructive solutions for roofs". (Source: authors).

Table 2: "Mean energy demands according to the position of the housing units". (Source: authors)

Position of the housing unit	**Heating Demand (kWh/m^2)**	**Cooling Demand (kWh/m^2)**	**Total Demand (kWh/m^2)**	**% number of housing units analysed**
Floor	39.6	20.1	59.8	8%
First Floor	38.8	20.6	59.6	6%
Intermediate	25.3	24.1	49.6	72%
Roof	40.6	32.2	72.7	14%

For the block model, the mean demand values obtained exceeded the limit stated in the current regulations, with heating demand values around 30 kWh/m^2 and cooling demand values around 24 kWh/m^2. It should be noted that the lowest total demands occurred in central positions and for H-shaped block typologies, while the most unfavourable occurred in isolated or corner positions in linear blocks (Table **3**).

Table 3: "Mean energy demands depending on block typology and position". (Source: authors)

Block Typology	Position of the block	Heating Demand (kWh/m^2)	Cooling Demand (kWh/m^2)	Total Demand (kWh/m^2)	% number of blocks analysed
Linear Block	Central	28.8	25.4	54.0	17%
	Corner	35.9	25.2	60.6	17%
	Isolated	37.3	23.3	61.4	9%
"H-shaped" Block	Central	29.0	22.2	51.2	23%
	Corner	34.6	23.5	58.6	18%
	Isolated	34.4	18.8	53.2	2%

In the energy evaluation of the original condition of the social housing units and housing blocks we found energy ratings from E to G. In the housing unit model in general there are mean energy ratings of G in the position of the floor, first floor and roof, while in the most numerous case, the 72% of housing units in intermediate positions, F predominates. In the case of the block model, the predominant rating for block corner and isolated positions is letter G, and for the central block position F predominates, which is always worse for the linear block typology than for the H-shaped block (Table **4**). The energy rating results obtained for the block model were represented on the plan of the city (Fig. **16**), thus offering a global vision of the energy conditions of social housing in the city of Cadiz, and clearly showing the major potential for improvement and energy saving, prior to the administration adopting efficient energy saving policies and subsequently selecting the most appropriate strategies for intervention.

Table 4: "Mean energy rating of block models depending on building typology". (Source: authors)

Position of block	Rating ($kgCO_2/m^2$ year)			
	Linear blocks		"H-shaped" blocks	
Central	F	28.2	F	26.5
Corner	G	30.3	G	29.1
Isolated	G	30.1	F	26.6

CONCLUSION

When applied to a city with Mediterranean climate, the methodology developed is an efficient tool for the formulation of sustainable planning policies, understanding the energy problem on a city scale with suitable typological, constructive and energy knowledge of the existing residential housing stock, prior to energy planning and intervention, and also forming the basis for the development of predictive models of energy performance.

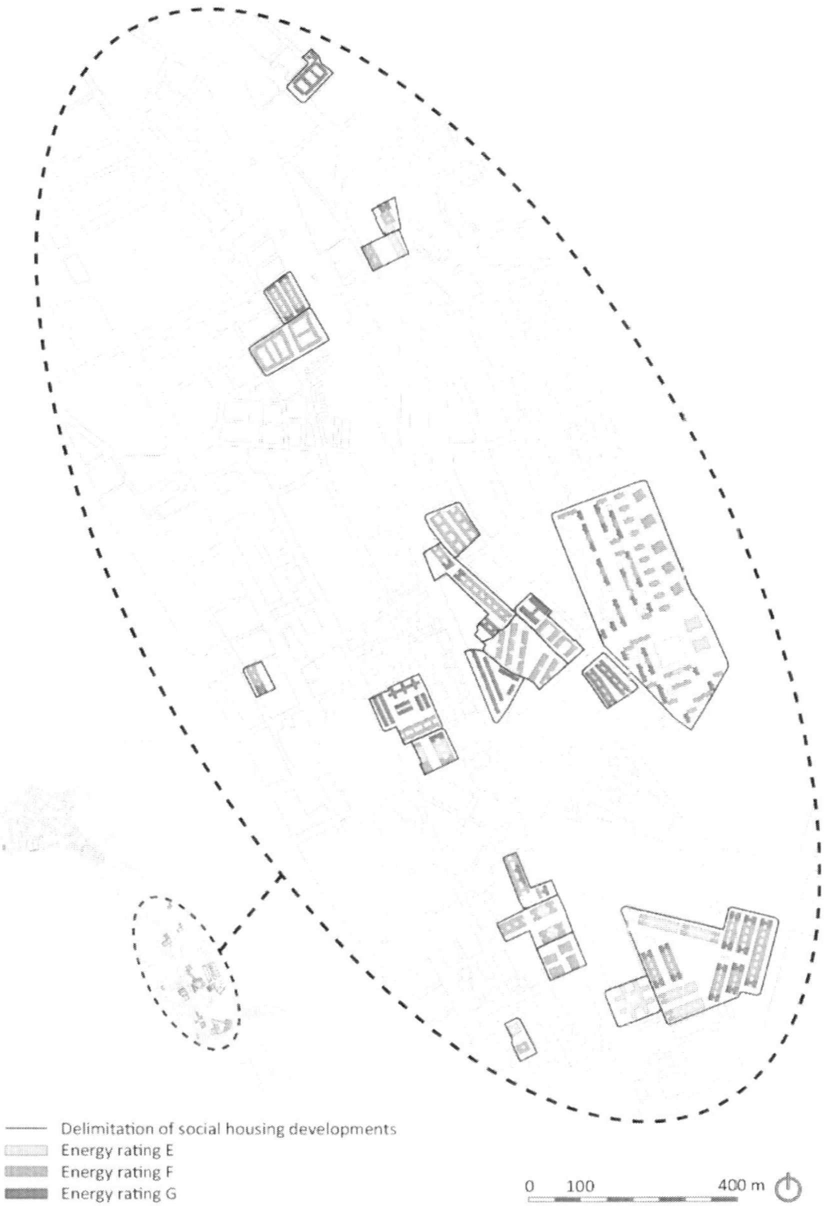

Figure 16: "Energy rating of the social housing developments analysed in their original condition in Cadiz". (Source: authors).

The city of Cadiz has a large number of multi-family social housing developments built between 1940 and 1980, with a growing increase observed in the number of housing units throughout the period analysed. The two main typologies are linear blocks and "H-shaped" blocks, mostly three-bedroom and with block height generally increasing over the period analysed, but with a clearly decreasing free height between floors, from 2.9-3 m in the 1940s to 2.5 m in the 1970s.

As regards the constructive systems used, there is a predominance of single-layer façades in the two first decades and subsequently two layers with air chamber, unventilated flat roofs and openings with metal joinery and single glazing. The thermal transmittance values for the envelope are much higher than the current regulation reference guidelines, due to the lack of thermal insulation, which translates into higher mean demand values than those established for climate zone A3, mainly heating.

Following the results of the energy ratings, we find that the energy condition in the city of Cadiz characteristically has very poor ratings in major residential neighbourhoods built between 1940 and 1980. This highlights the need to develop a city scale procedure for evaluating the potential for improvement and energy saving in order to help to attain the current priority action aims of Horizon 2020, through global actions with integral sustainable retrofitting plans, focusing on the main solutions for retrofitting of the thermal envelopes.

ACKNOWLEDGEMENTS

This work has been financially supported by the Spanish Ministry of Science an Innovation, with reference BIA2012-39020-C02-01.

CONFLICT OF INTEREST

The authors confirm that this chapter contents have no conflict of interest.

REFERENCES

[1] EC, "Financial support for energy efficiency in buildings, Report from the Commission to the European Parliament and the Council", 2013. [Online] Available; http://ec.europa.eu/energy/efficiency/buildings/doc/report_financing_ee_buildings_com_2013_225_en.pdf

[2] BPIE, "Europe's buildings under the microscope, A country-by-country review of the energy performance of buildings", 2011. [Online] Available; http://bpie.eu/documents/BPIE/WEB_executive%20summary.pdf (Last visit: 17-04-2015)

[3] European Commission, "*Energy Roadmap 2050*", Brussels, COM, 885/2, 15 December 2011. [Online] Available; http://ec.europa.eu/energy/energy2020/roadmap/doc/com_2011_8852_en.pdf.

[4] R. Galvin, M. Sunikka-Blank, "Economic Viability in Thermal Retrofit Policies: Learning from Ten Years of Experience in Germany", *Energy Policy*, vol. 54, pp. 343-351, 2013.

[5] I.G. Hamilton, Philip J. Steadman, H. Bruhns, A. J. Summerfield, R. Lowe. "Energy efficiency in the British housing stock: Energy demand and the Homes Energy Efficiency Database", *Energy Policy*, vol. 60, pp-462-480, 2013.

[6] J.J. Sendra, S. Domínguez, P. Bustamante, A.L. León, "Energy Intervention in the residential sector in the south of Spain: Current challenges", *Informes de la Construcción*, vol. 65, 532, pp- 457-464, 2013.

[7] E. Dascalaki, K. G. Droutsa, A. C. Balaras , S. Kontoyiannidis. "Building typologies as a tool for assessing the energy performance of residential buildings. A case study for the Hellenic building stock", *Energy and Buildings*, vol. 43 (12), pp- 3400-3409, 2011.

[8] G. Dall'O, A. Galante, G. Pasetti. "A methodology for evaluating the potential energy savings of retrofitting residential building stocks", *Sustainable Cities and Society*, vol. 4, pp- 12-21, 2012.

[9] G. Dall'O', A. Galante, M. Torri, "A methodology for the energy performance classification of residential building stock on an urban scale", *Energy and Buildings*, vol. 48, pp. 211-219, 2012.

[10] G. Dall'O', M.F. Norese, A. Galante, C. Novello, "A multi-criteria methodology to support Public Administration decision making concerning sustainable energy action plans", *Energies*, vol 6, pp. 4308-4330, 2013.

[11] A. Mastrucci, O. Baume, F. Stazi, U. Leopold, "Estimating energy savings for the residential building stock of an entire city: A GIS-based statistical downscaling approach applied to Rotterdam", *Energy* and Buildings, vol. 75, pp. 358-367, 2014.

[12] G.V.Fracastoro, M. Serraino, "A methodology for assessing the energy performance of large scale building stocks and possible applications", *Energy and Buildings*, vol. 43, pp. 844-852, 2011.

[13] V. Belpoliti, G. Bizzarri, "A parametric method to assess the energy performance of the social housing stock and simulate suitable retrofit scenarios: An Italian case study", *Energy and Buildings*, vol. 96, pp. 261-271, 2015.

[14] G.P. Panayiotou, S.A. Kalogirou, G.A. Florides, C.N. Maxoulis, A.M. Papadopoulos, M. Neopfytou, P. Fokaides, G. Georgiou, A. Symeou, "The characteristics and the energy behaviour of the residential building stock of Cyprus in view of Directive 2002/91/EC", *Energy and Buildings*, vol. 42, pp. 2083-2089, 2010.

[15] I. Theodoridou, A.M. Papadopoulos, M. Heggera, "Statistical analysis of the Greek residential building stock", *Energy and Buildings*, vol. 43, pp. 2422-2428, 2011.

[16] L.G. Swan, V.I. Ugursal, "Modeling of end-use energy consumption in the residential sector: A review of modeling techniques", *Renewable and Sustainable Energy Reviews*, vol. 13, pp. 1819-1835, 2009.

[17] F. J. S. De la Flor, S. A. Domínguez, J. L. M. Félix, R. G. Falcón, "Climatic zoning and its application to Spanish building energy performance regulations", *Energy and Buildings*, vol. 40, pp. 1984-1990, 2008.

[18] Sede Electrónica de la Dirección General del Catastro. Date accessed: March 2014. [Online]: http://www.sedecatastro.gob.es/

[19] "*Royal Decree 235/2013, approving the Basic Procedure for the Certification of Energy Efficiency in Buildings*". Sec. I. p. 27548, 2013.

[20] "Order FOM/1635/2013, of 10 September, updating DB-HE *«Ahorro De Energía», of the Building Technical Code*", approved in Royal Decree 314/2006, of 17 March. Sec. I. P. 67137, 2013".

Chapter 9: New Applications of Ceramic Waste Generated in Building Rehabilitation Works

CHAPTER 9

New Applications of Ceramic Waste Generated in Building Rehabilitation Works

Mercedes Del Río Merino*, Rocío Sanios Jiménez, Paola Villoria Saéz and Mariano Gonzalez Cortina

ETSEM-UPM, Department of Building Construction and its Control, Juan de Herrera 6 Avenue, 28040, Madrid, Spain

Abstract: The European Union has identified construction and demolition waste (CDW) flow as a priority waste stream. In 2012, more than 26 million tons were generated in Spain, of which almost 33% was derived from the building sector. Besides the strong growth in the building activity since 1991-2011 (20.3% *versus* population growth was only 14%) focused on newly built constructions, in the past recent years, a new paradigm is currently rising in Spain, changing from new construction activities to building rehabilitation works, mainly due to the severe economic crisis. This change together with the recent legal requirements for building Energy Efficiency will lead to a considerable increase of building rehabilitation and, in particular, energy retrofitting works, pushing the construction sector towards a new challenge. This new challenge will generate vast amounts of CDW, as building rehabilitation works generate higher amounts of waste compared to newly built constructions. In addition, according to the I National Waste Management Plan published by the Ministry of Environment, 54% of the total CDW generated in Spain corresponds to ceramic materials. This situation has led to the definition of research seeking the recovery of these wastes in order to promote their reduction and avoid landfilling, promoting as well the correct waste management to turn them into resources and save raw materials. Therefore, this chapter presents the results of a research project conducted to analyze the feasibility of incorporating ceramic brick and tile waste from construction in a plaster matrix, through a study carried out in laboratory and ensuring meeting the current regulations. Results show that it is possible to incorporate up to 25% of ceramic waste over the weight of gypsum without changing its basic properties. Perforated brick waste achieves the best behavior: increases mechanical strengths (31% compression and 29% flexural) and surface hardness (8%), and reduces water absorption around 25%.

Keywords: CDW, recycling, construction materials, ceramic, plaster.

INTRODUCTION

The European Union has identified the flow of construction and demolition waste (CDW) as a priority waste stream. In 2012, Spain generated around 26 million

***Corresponding author Mercedes Del Río Merino:** Department of Building Construction and its Control, ETSEM-UPM, Madrid, Spain; Tel/Fax: +34 91 3367646; E-mail: mercedes.delrio@upm.es

Pilar Mercader-Moyano (Ed.)

tonnes of CDW [1]. CDW is mainly produced during demolition and construction activities, and it is classified under chapter 17 in the European Waste List (Order MAM / 304/2002). In order to reduce the amount of CDW generated several countries are implementing regulations to prevent the waste generation and to promote recycling measures, as for example, the Royal Decree 105/2008 in Spain [2]. This Law, forces to draw up a Waste Management Plan and Report for each building construction project.

Despite the high potential of CDW recovery, still construction agents continue to choose elimination *vs*. recycling. According to the 2nd National Plan of CDW 2007-2015 (II PNCDW) Spain recycled around 18% of the CDW generated [3]. This situation is mainly due to the type of waste collection system, normally managed by each subcontracted company in the construction work [2]. Therefore, implementing measures for reducing CDW is not a common practice in the construction works, as it is not considered during the design stage. Therefore, there is an important lack of planning on-site best practices to minimize waste and promote recycling. Nevertheless, main construction companies are slowly considering environmental issues by implementing Environmental Management Systems (EMS), not being limited to the current legislation, and seeking the implementation of good environmental practices in their works [4]. However, while this fact is a reality for large construction companies, still the vast majority of construction companies (small and medium enterprises) need to accept this trend [5].

Moreover, it is important to highlight two issues that are changing the scenario of the Spanish building sector. On the one hand, the serious crisis that the sector has suffered in the last five years mainly due to the huge unsold properties stock and on the second hand the government measures aimed at improving the energy efficiency of buildings, as for example the Technical Building Code --basic document "Energy Saving"--. In this sense the building sector has changed its paradigm towards building rehabilitation works and, in particular, building energy efficiency works, as a solution to the crisis. This change has caused a significant increase of rehabilitation works in Spain, and therefore CDW generation rates are recovering values similar to the beginning of the crisis, as CDW ratio (CDW kg per m^2 rehabilitation/construction) is much higher than new construction (338.7 kg/m^2 and 120.0 kg/m^2 respectively) [3].

Despite the extensive bibliography considering CDW management for new constructions [6] – quantification methods [7-9] and good practices [4, 10, 11] -- few references were found considering building rehabilitation works. In this sense, several researchers from different Schools of Building Engineering and Architecture in Spain are developing a research project called "W2R" aiming to quantify the CDW generated in building retrofitting works and analyze their feasibility to be used as raw materials for the production of new composites in order to design low-energy-consumption solutions for the building envelope. The main objectives of the project are:

1. Classification of Spanish existing buildings needing energy efficiency retrofitting. Quantification of the waste generated.

2. Feasibility of recycling CDW as raw materials for the manufacture of new construction materials.

3. Development and evaluation of new construction elements and solutions for the building envelope using these new materials with CDW.

This article presents some of the first results of the second phase. In particular, the feasibility of incorporating ceramic waste in gypsum based materials manufacture in order to improve some of their benefits.

Construction and Demolition Waste Generated by Ceramic Elements

In general, the greatest percentage of CDW generated in Europe, for both new constructions and rehabilitation works, corresponds to masonry and concrete waste [7]. According to del Río *et al.* (2011) study approximately 60% of masonry waste is generated during the demolition stage [2]. Masonry waste characteristics can vary significantly depending on different factors, such as type of structure, main masonry elements (brick, concrete, *etc.*). Masonry waste can be easily recycled and used as aggregates for building materials. However, only 6% of the aggregate production was recycled aggregates. This low percentage points out the need to increase the use of recycled aggregates in building materials. However, one of the main problems of using recycled CDW aggregates is that manufacturers have shown little interest in using them as raw materials, maybe due to the low number of regulations considering CDW as raw materials. In Spain, only the Structural Concrete Instruction EHE-2008 provides in its Annex 15 recommendations for the use of recycled aggregates for manufacturing concrete [12].

This situation has promoted the publication of different manuals about recycling waste [13-15]. However, the list of materials considered in these manuals can be increased, as no manual was found establishing strategies for recycling wastes in gypsum composites. Therefore, several researchers have been concerned about this situation, conducting studies to analyse the feasibility of incorporating CDW in building materials [16-19]. In particular, the following waste was included in gypsum composites: paper [20], expanded polystyrene [21], polyurethane foam [22], rubber [23], *etc.* All these studies were focused on gypsum reinforcement, increasing its surface hardness or decreasing its density.

Moreover, ceramic wastes have been also studied in lime mortars [24], alkaline cements [25] or cement mortars [26], but none of these studies indicate and verify what kind of applications can be given to such materials. Besides, several authors have investigated the incorporation of ceramic waste in concrete [27] and its application in precast concrete [28], there are no studies dealing with the feasibility of incorporating such waste in gypsum composites.

OBJECTIVE

The aim of this study is to analyse the feasibility of incorporating ceramic waste as an additive to increase the surface hardness and reduce water absorption by capillarity in a gypsum matrix because they are common problems that occur in coatings or prefabricated plaster.

MATERIALS AND METHODS

Gypsum plaster E-30 according to the European Standard EN 13279-1 was used. Initially the test composites were selected according to particle size and percentage of ceramic waste addition. Moreover, two types of ceramic waste from building works were used: perforated bricks (test A), and flat tiles (test B). For each ceramic waste, different composites were selected according to the following variables: particle size (coarse or fine aggregate) and percentage of added waste (25%, 50% and 75% of gypsum weight). A total of 12 composites and 2 reference samples (REF) were developed and the water/binder (W/B) ratio for each sample was determined according to European Standard EN 13279-2 (Table **1**).

Table 1: List of composites

	Composite	Binder	W/B ratio	% of waste added by gypsum weight	Aggregate
Test A	REF	E-30	0.70	-	-
	25AG	E-30	0.60	25%	AG
	50AG	E-30	0.70	50%	AG
	75AG	E-30	0.80	75%	AG

Table 1: contd….

	25AF	E-30	0.70	25%	AF
	50AF	E-30	0.82	50%	AF
	75AF	E-30	0.90	75%	AF
Test B	REF	E-30	0.70	-	-
	25BG	E-30	0.67	25%	BG
	50BG	E-30	0.80	50%	BG
	75BG	E-30	0.90	75%	BG
	25BF	E-30	1.00	25%	AF
	50BF	E-30	1.15	50%	AF
	75BF	E-30	1.20	75%	AF

AG: Perforated brick waste (coarse aggregate: 60% of ø2mm and 40% of ø1mm).
AF: Perforated brick waste (fine aggregate: 25% of ø1mm, 25% of ø0.5mm, 25% of ø0.25mm and 25% of ø<0.25mm).
BG: Flat tile waste (coarse aggregate: 60% of ø2mm and 40% of ø1mm).
BF: Flat tile waste (fine aggregate: 25% of ø1mm, 25% of ø0.5mm, 25% of ø0.25mm and 25% of ø<0.25mm).

Further, physical and mechanical properties of each composite were determined by testing prismatic specimens (160x40x40) mm^3 according to European Standard EN 13279-2. All samples were characterized by their dry density, mechanical strength at 7 days, surface hardness and capillary water absorption. Finally, all the results were compared to the values obtained with the reference sample (without ceramic waste).

RESULTS AND DISCUSSION

Results are shown depending on the type of ceramic waste used:

- Test A: made with plaster E-30 as a base and additions of ceramic waste from perforated bricks of new building work.

- Test B: made with plaster E-30 as a base and additions of ceramic waste from flat tile of new building work.

Results from gypsum composite with perforated brick waste (test A) are shown in Fig. **1** and Fig. **2**.

The best results are obtained with composite 25AG increasing around 21% its density when compared to the reference value. Also, an increase in flexural strength (29%), compression strength (31%), and Shore C surface hardness (8%) was achieved with this composite; despite reducing capillary water absorption around 25%.

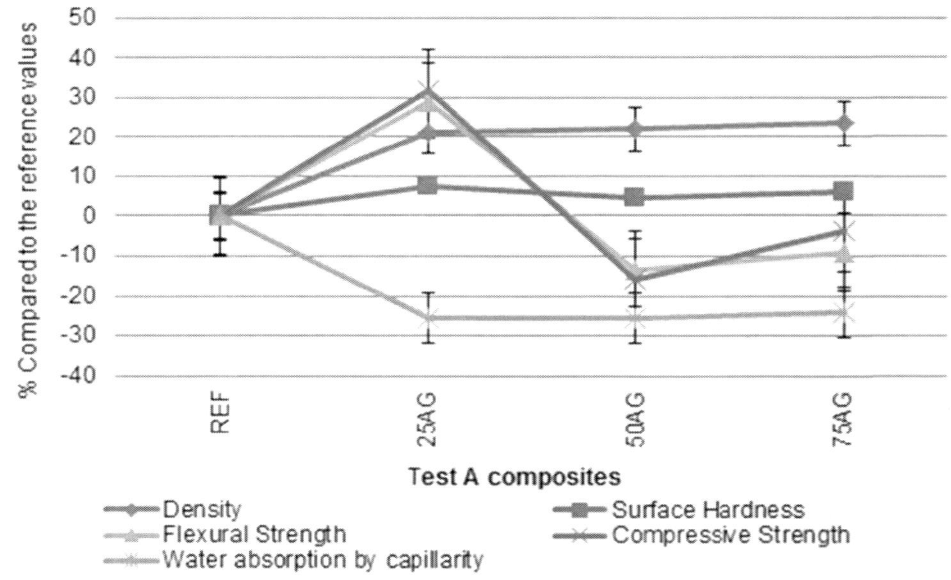

Figure 1: Results from gypsum composites with coarse aggregates of perforated bricks waste (A).

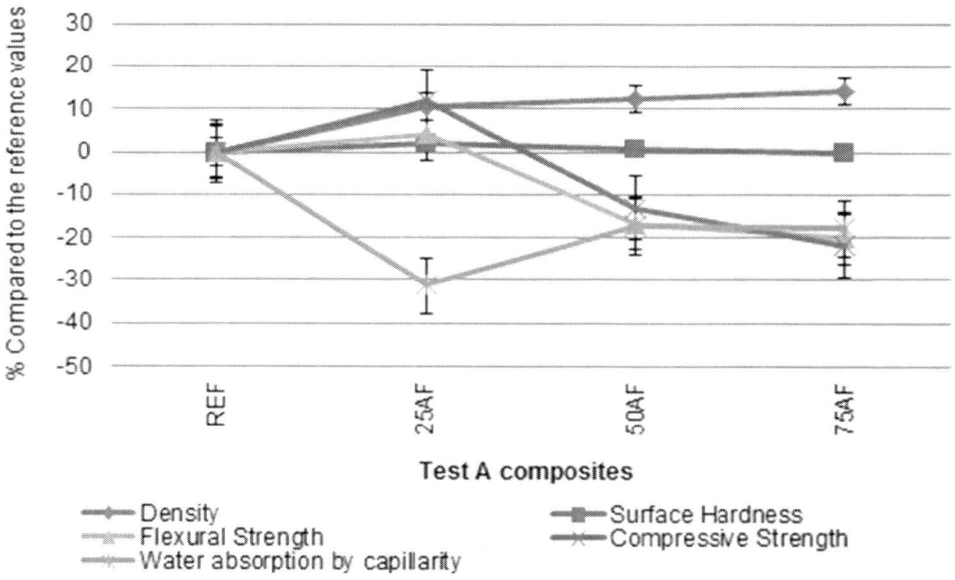

Figure 2: Results from gypsum composites with fine aggregates of perforated bricks waste (A).

According to the results obtained for the composites with additions of ceramic waste from flat tiles (test B), both coarse aggregate and fine aggregate, compared to the reference sample, are shown in Fig. **3** and Fig. **4**.

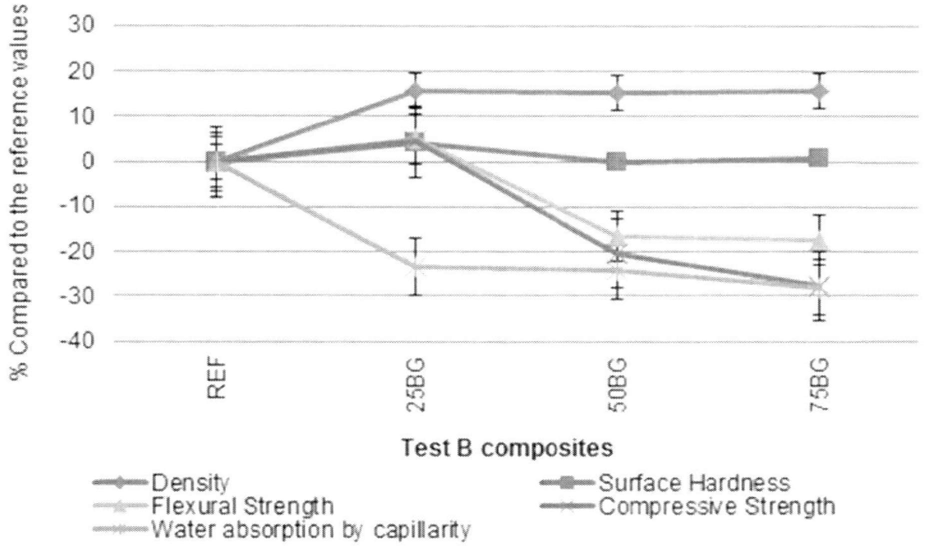

Figure 3: Results from gypsum composites with coarse aggregates of ceramic flat tile waste (B).

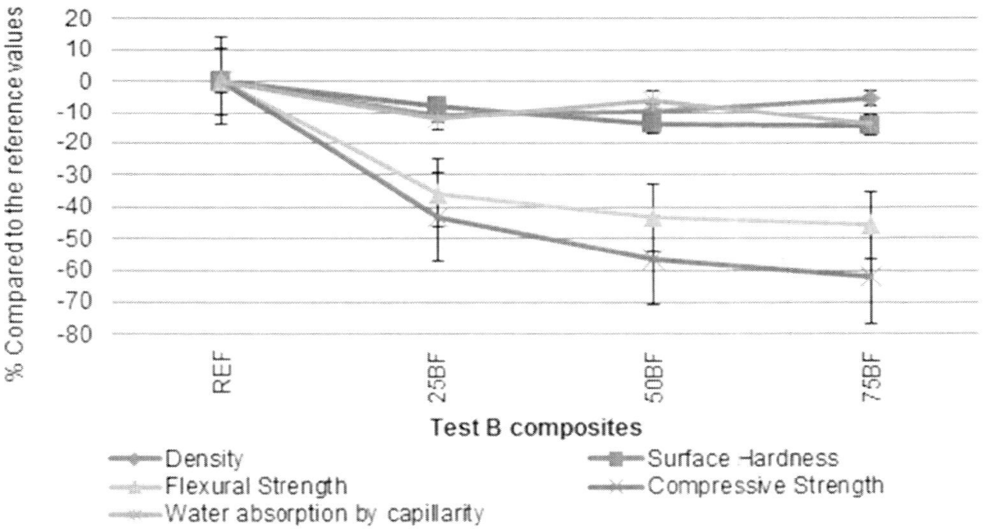

Figure 4: Results from gypsum composites with fine aggregates of ceramic flat tile waste (B).

From the results obtained composite 25BG showed an increase in: density around 15%, flexural and compression strength around 5%, and Shore C surface hardness about 5%; whereas capillary water absorption was reduced 24% compared to the reference value.

Additionally, Table **2** summarizes the best composites obtained for either test A and B, showing possible applications in the construction industry.

Table 2: Best composites indicating their physical-mechanical behaviour and possible applications

Best composite	Type of waste	Particle size/ added waste		Composites behavior in comparison to a composite without additions (%)					Possible applications
		ø mm	%*	Density	Superficial hardness	Flexural Strength	Compressive Strength	Capillary water absorption	
25AG	Perforated brick	2.0	15%	+21%	+8%	+29%	+31%	-25%	Prefabricated elements Interior coatings
		1.0	10%						
25BG	Flat tile	2.0	15%	+15%	+5%	+5%	+5%	-24%	Prefabricated elements Interior coatings
		1.0	10%						

*Percentage of added waste by plaster weight.

Despite the increase of density, which can affect materials' transport, both composites highly improve both flexural and compressive strengths and superficial hardness, and reduce water absorption. Therefore, the composite is suitable to be used for prefabricated elements manufacture or as interior finishing material. All composites meet the requirements set by the current European Standard regulations for gypsum plaster E-30, as results surpass the minimum flexural and compression strength (3.0 N/mm^2 and 2.0 N/mm^2 respectively).

CONCLUSION

From the analysis performed and the results obtained, it is feasible to add 25% of coarse aggregate ceramic waste --both perforated brick and flat tile waste-- in a gypsum plaster E-30 matrix, fulfilling the minimum requirements established by the current legislation. Flexural and compression strengths can be increased around 30% --compared to the reference sample-- with perforated brick waste, and only 5% with ceramic flat tile waste. Also the surface hardness increases around 8% and the water absorption decreases around 25%. Therefore, this composite can be used for the manufacture of prefabricated gypsum elements and also as material for interior coatings. Finally, this analysis can be enhanced by studying other factors such as setting time and adherence or thermal conductivity.

ACKNOWLEDGEMENTS

This study has been carried out as part of the BIA2013-43061-R research project, funded by the Spanish Ministry of Economy and Competitiveness (The State Secretariat for Research, Development and Innovation).

CONFLICT OF INTEREST

The authors confirm that this chapter contents have no conflict of interest.

REFERENCES

[1] European Commision. (2014, December). Eurostat statistics for waste flow generation 2012. Available: http://epp.eurostat.ec.europa.eu/portal/page/portal/eurostat/home/

[2] M. del Río Merino, P.I. Gracia, I.S.W. Azevedo. "Sustainable construction: construction and demolition waste reconsidered", *Waste Management & Research*, vol. 28, pp. 118-129, 2010.

[3] Spanish government, "II Plan Nacional de Residuos de Construcción y Demolición (2008-2015). "II National Plan for Construction and Demolition Waste (2008-2015)" ed. Official state bulletin (BOE): *Ministry of Environment*, 2008, pp. 19938-19948.

[4] P. Villoria Saez, *et al.*, "Best practice measures assessment for construction and demolition waste management in building constructions," *Resources, Conservation and Recycling*, vol. 75, pp. 52-62, 2013.

[5] M. del Río Merino, *et al.*, "Los sistemas de gestión de la calidad en la edificación instrumentos para la reducción del impacto medioambiental," In: *Actas del I Congreso Internacional de Construcción Sostenible y Soluciones Ecoeficientes.*, Sevilla 2013, pp. 74-84.

[6] H. Yuan and L. Shen, "Trend of the research on construction and demolition waste management," *Waste Management*, vol. 31, pp. 670-679 2011.

[7] M. Mália, J. de Brito, M.D. Pinheiro, M. Bravo. "Construction and demolition waste indicators," *Waste Management & Research*, vol. 31, pp. 241-255 2013.

[8] P. Villoria Saez, M. del Río Merino, C. Porras-Amores. "Estimation of construction and demolition waste volume generation in new residential buildings in Spain," *Waste Management & Research*, vol. 30, pp. 137-146, FEB 2012 2012.

[9] M. P. Mercader-Moyano and A. Ramirez-de-Arellano-Agudo, "Selective classification and quantification model of C&D waste from material resources consumed in residential building construction," *Waste Management & Research*, vol. 31, pp. 458-474, May 2013.

[10] W. Lu and H. Yuan, "Exploring critical success factors for waste management in construction projects of China," *Resources, Conservation and Recycling*, vol. 55, pp. 201-208 2010.

[11] H. Yuan, "Critical Management Measures Contributing to Construction Waste Management: Evidence From Construction Projects in China," *Project Management Journal*, vol. 44, pp. 101-112, Aug 2013.

[12] Spanish Government, "Structural concrete regulation (EHE: Instrucción de hormigón estructural)," ed: *Ministerio de Fomento*, 2008.

[13] CEDEX, "Catalogue of construction and demolition waste that can be used in construction (Catálogo de residuos que pueden utilizarse en construcción)," ed. Madrid, 2001.

[14] Australian Government, "Construction and demolition waste guide recycling and re-use across the supply chain.," Australia 2012.

[15] Gruzen Samton LLP and City Green Inc, "*Construction and demolition waste manual*," New York City2003.

[16] M. Marrero, L. Martínez-Escobar, M.P. Mercader, C. Leiva. "Building facades: environmental impact minimization by means of recycled material usage," *Informes de la Construcción*, vol. 65, pp. 89-97, 2012.

[17] C. Leiva, L.F. Vilches, J. Vale, C. Fernández-Pereira. " Fire-resistance of biomass ash panels used for internal partitions in buildings.," *Fire Safety Journal*, vol. 44, pp. 622 - 628, 2009.

[18] D. Alba, M. Marrero, C. Leiva, M. V. Montes, L. Vilches. "Facade solutions using panels made of power plant byproducts.," *Informes de la Construcción*, vol. 64 pp. 179-190, 2012.

[19] A. San-Antonio-Gonzalez, R.S. Jimenez, M. del Rio Merino, M.G.C.C.V. Arrebola. "Feasibility of Recycling CDW as Raw Materia in Gypsum Composites.," *Athens Journal of Technology & Engineering* (AJTE), January, 2015.

[20] W. Klock and S. Aicher, "Size effect in paper fiber-reinforced gypsum panels under in-plane bending," *Wood and Fiber Science*, 2005.

[21] F. J. González Madariaga and J. Lloveras Macia, "Mezclas de residuos de poliestireno expandido (EPS) conglomerados con yeso o escayola para su uso en la construcción," *Informes de la Construcción*, vol. 60, pp. 35-43, 2008, Vol. 28, no.1, pp. 653-658.

[22] S. Gutierrez-Gonzalez, J. Gadea, A. Rodríguez, C. Junco, V. Calderón. "Lightweight plaster materials with enhanced thermal properties made with polyurethane foam wastes," *Construction and Building Materials,* 2012.

[23] A. Jiménez Rivero, A. de Guzman Baez, J. G. Navarro. "New composite gypsum plaster – ground waste rubber coming from pipe foam insulation," *Construction and Building Materials*, vol. 55, pp. 146-152, 2014.

[24] M. González cortina and L. Villanueva domínguez, "*Recuperación de morteros romanos de cal y chamota en aplicaciones actuales*" Tesis Doctoral, Escuela Técnica Superior de Arquitectura. Departamento de Construcción y Tecnología Arquitectónicas, Universidad Politécnica de Madrid, Madrid, 2000.

[25] F. B. Puertas, A; Gazulla, MF; Gómez, MP; Palacio, M; Martínez-Ramírez, S., "Ceramic wastes as raw materials in portland cement clinker fabrication: Characterization and alcaline activation" *Materiales de Construcción*, vol. 56, pp. 73-84, 2006.

[26] M. S. Chiara Bignozzi, A., "Ceramic waste as aggregate and supplementary cementing material: A combined action to contrast alkali silica reaction (ASR)" *Cement & Concrete Composites*, vol. 34, pp. 1141-1148, 2011.

[27] C. Medina, A. Juan, M. Frías, M. I. Sánchez de Rojas, J. M. Morán, M. I. Guerra. "Characterization of concrete made with recycled aggregate from ceramic sanitary ware," *Materiales de Construcción*, vol. 61, pp. 533-546, 2011.

[28] M. Sánchez de Rojas, F.P. Marín, M. Frías, J. Rivera. "Viability of utilization of waste materials from ceramic products in precast concretes," *Materiales de Construcción*, vol. 51, pp. 263-264, 2001.

Author Index

Subject Index

Subject Index

Pilar Mercader-Moyano (Ed.)

Made in United States
North Haven, CT
04 November 2026

21719585R00111